MAPPING LONDON
Making Sense of the City

Simon Foxell

black dog
publishing

To Imogen, Mischa and Pascoe who are also Londoners

Contents

INTRODUCTION

In the Elliot Avedon Museum and Archive of Games in Waterloo, Ontario, Canada there is a folded board entitled, The Road to Wealth: How to Know London. Any other parts, and the instructions that went with it, are missing. No one knows exactly how this early nineteenth century game was played, but it was clearly done so by moving pieces across a map of London. The Thames, with its characteristic wriggles, snakes from west to east across the board, while a number of roads and parks are marked. The sites of various landmarks, listed in the title box, are indicated by rectangles, suggesting that the players had to position appropriate pieces on them as the game proceeded. The game poses the question: "How to Know London", and the answer is clear enough; by consulting a map, losing oneself amongst the lines and shapes that represent the city and beginning to understand the patterns, flows and intricate relationships that have been traced by the cartographer.

The first part of the title, "The Road to Wealth", is more opaque. Is familiarity with London's roads and landmarks the pathway to wealth? Not since Henry Dodd, 1801–1881—the ploughboy turned rubbish collector, and the likely model for Dickens' Nicodemus Boffin with his 'golden' mounds of dust in *Our Mutual Friend*, 1865—has much wealth been gathered from, rather than displayed on, the streets of London. Maps of and mapmaking in London have many other hidden treasures, however, promises of great wealth are probably not among them.

But knowledge is there, even if understanding has to be unearthed. Any map potentially includes all human knowledge and much that is potent but has yet to be understood. The map, as a scaled replica of the entire city, presents a choice to its maker: not what to include, but rather, what to exclude. The mapmaker, like a sculptor, must chip away at the raw block of material that is the city to reveal the shape and representation hidden inside. Over the centuries, that raw block has revealed the most astounding shapes and interpretations under the hands of some of the most talented, imaginative and creative individuals. Carving away, they have revealed a London that we didn't know existed or that we needed to find. The cartographers of London have made sense of the city in ways that compare with writers, such as William Chaucer through to Martin Amis, who have animated London with their characters and narratives: "There was a time when I thought I could read the streets of London. I thought I could peer into the ramps and passages, into the smoky dispositions, and make some sense of things. But now I don't think I can. Either I'm losing it, or the streets are getting harder to read. Or both."[1]

Yet mapmaking is a very different means of revealing the underlying patterns of the city. A map contains layer upon layer of information, enabling intricate connections to be recognised and made—yet it can be scanned and an initial understanding reached in an instant. Much time can, of course, be spent exploring detail, getting lost in the underlying complexity and richness of its links and relationships. Maps, and the diagrams that are their close cousins, are the best understood means of graphical communication and, quite possibly, the most powerful way of conveying detailed information in a clear form. Maps are instinctively understood by a large proportion of the population. They are ideal tools, despite being partially word-based, for explaining the complexities of London to non-English speaking strangers. Yet there are still many that find maps opaque representations and the stylised conventions of the cartographer as difficult to read as a foreign language.

Maps provide ways of understanding physical constructs, human society, narrative complexities and other such subjects that are far from readily apparent in the day-to-day navigational diagrams familiar to everyone living in the modern world. For, as Stephen Hall notes in his introduction to *You Are Here: Personal Geographies*, "out of one territory, one map can bloom a thousand geographies".[2] The mapmakers whose maps are shown and discussed in this book are explorers who have been intent on discovering those geographies on our behalf, who have crossed *terra incognito* in pursuit of the dragons that dwell there. On their return, they have told unlikely stories of their findings and adventures; they have been canonised in the manner of returning heroes but also laughed at, mocked, or worse; ignored.

To tell their stories, and to reveal the information painstakingly gathered in their investigations, mapmakers have developed a myriad of new approaches. In the course of their mapmaking, the technologies available have developed in exciting and innovative ways, from the invention of printing to Global Positioning Systems (GPS) in common use today. Maps have been made to do things that they were never expected to tackle and, in doing so, have exposed patterns, connections and ideas that were as interesting as they were unexpected.

The straightforward (to our minds) geographical map that describes the lay of the land has become a snapshot in time, showing not only how the city was during a specific period, but also the issues that were uppermost in the mapmaker's consciousness.

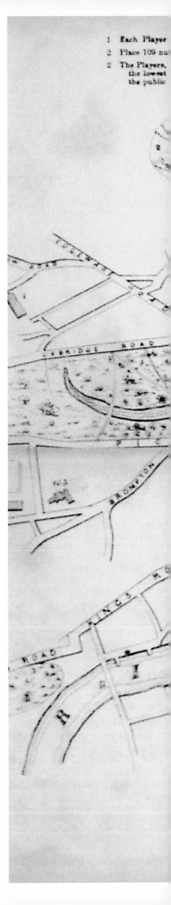

The Road to Wealth—How to Know London,
circa 1860

Image courtesy of Elliott Avedon Museum and Archive
of Games, University of Waterloo, Canada.

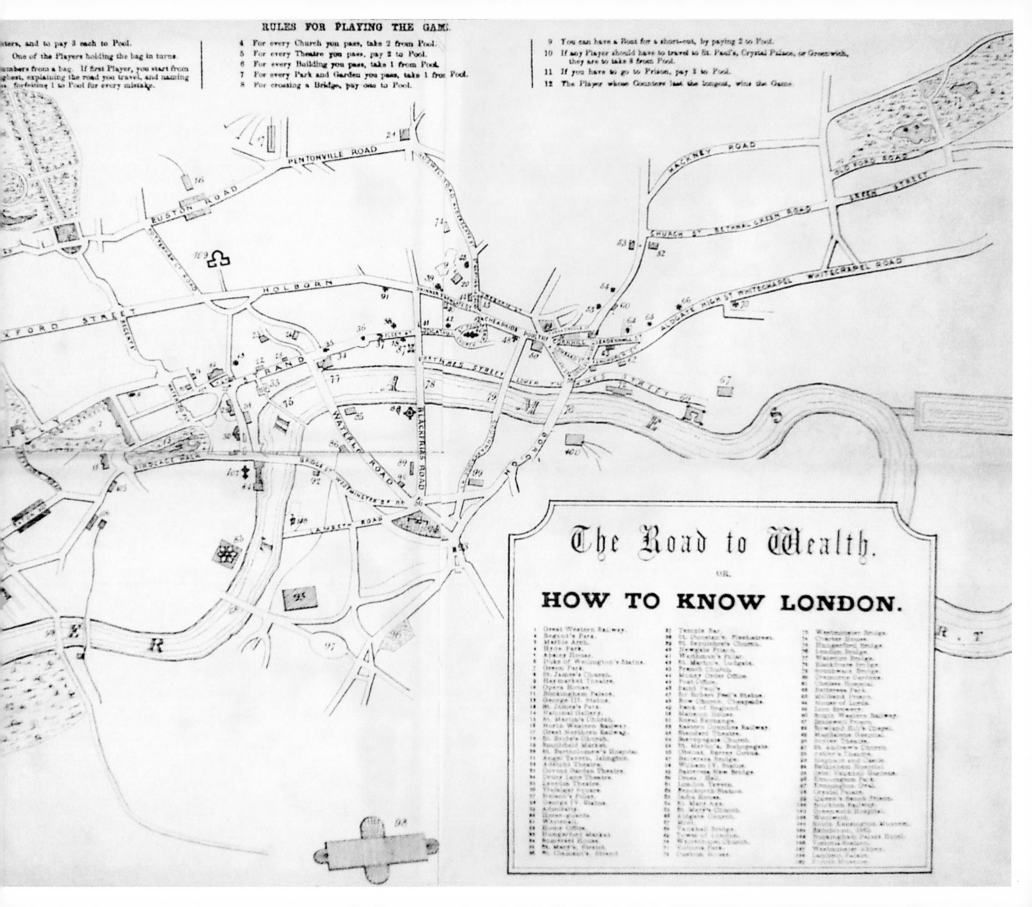

RULES FOR PLAYING THE GAME.

...ters, and to pay 3 each to Pool.

One of the Players holding the bag in turns.

...umbers from a bag. If first Player, you start from ...ghest, explaining the road you travel, and naming ... forfeiting 1 to Pool for every mistake.

4 For every Church you pass, take 2 from Pool.
5 For every Theatre you pass, pay 2 to Pool.
6 For every Building you pass, take 1 from Pool.
7 For every Park and Garden you pass, take 1 from Pool.
8 For crossing a Bridge, pay one to Pool.

9 You can have a Boat for a short-cut, by paying 2 to Pool.
10 If any Player should have to travel to St. Paul's, Crystal Palace, or Greenwich, they are to take 3 from Pool.
11 If you have to go to Prison, pay 3 to Pool.
12 The Player whose Counters last the longest, wins the Game.

The Road to Wealth.
OR.
HOW TO KNOW LONDON.

Like DNA, strands can be followed from one generation to the next, as received wisdom was accepted blindly in the face of evidence on the ground, before being crushingly rejected, and as conventions were adopted and standardised, and then mutated to become capable of new tasks.

The latest maps will become historical artefacts in their turn. The technology that allows us to call up maps at will or be guided through the streets by satellite navigation will be as revealing of our priorities as any map of Stuart, Georgian or Victorian London.

This book is divided into approximately themed sections that look at the variety of drawn geographies of London. They sketch the strata of London, starting with the historical record provided by mapmakers as the city stretched and drew itself into its current shape. They examine the way London has functioned and been governed, and provide 'morning-after' reports of the city's great parties and events. They also present re-interpretative imaginings of the contemporary city and how it might take form in the future.

As ever, the predominant use of maps in London is to help us to navigate—whether by foot, bike, car, bus or train. Millions of such maps are printed annually in book, folded or chart form. Many more are accessed on computer screens with just a small proportion finding their way into paper form.

The lives of all city dwellers, and visitors, are improved immeasurably by their use of maps—whether virtual or physical—to plot journeys through the maze of streets and train connections. Due to the needs of Londoners, and the sheer numbers of such maps produced and consumed, innovation in mapping the city has been restless and relentless, with the invention of the London Underground diagram by Harry Beck, in particular, marking out London as a city of cartographic genius. Competitive designers attempting to surpass Beck's plan have kept the navigational map thriving as the pre-eminent map form, yet the most significant innovation has lain elsewhere.

This must be the area of Geographical Information Systems (GIS), a branch of information science that has developed into map data, which shows spatial distributions and reveals visual linkages. London has a founding role in the development of GIS, starting with John Snow's map of the 1854 cholera outbreak in central London, from which it has developed as a tool used not only in health and epidemiology but also in the analysis of demography, criminology, planning, logistics, commerce and disaster management, among many other fields. GIS is still in its infancy, but its power is fully recognised and it is employed by a number of London organisations at all sorts of levels.

One of the least developed aspects of GIS is its predictive potential, for years the sole territory of meteorological studies. Mapping techniques are developing to deal with the more extreme environmental changes associated with climate change and the increasing likelihood of disastrous floods. They are also being used to predict accidents and crime and will become an essential tool in planning for the future as the ability to computer-model ever more complex events increases. In not so many years it should even be possible to model a city as multifarious and obtuse as London.

Prediction is different from planning, despite much common territory. Planning takes an imaginative leap into the future, with the intention of deliberately shaping events, and maps, one of its essential tools. The freedom of drawing on paper or, these days, on the computer, as opposed to building on the ground, allows all sorts of fantasies to be considered, enjoyed, manipulated and—in all likelihood—rejected. This freedom has generated mad, seductive and dangerous plans for London as well as concepts of great sense and beauty, although none has ever been able to compete with the high-level anarchy that has generally characterised the city's development. There are many plans for the city that, perhaps fortunately, never got further than paper—remaining only in map or diagram form. Some proposals and grand schemes managed to start work on the ground only to be abandoned and to leave a few buildings and streets in their wake. These, often utopian, fragments have made their mark on the map of London: traces of the many minds and generations who have tried to shape and plan it.

The imagination of London's planners has opened the door to others who have made free with the city's topography and who use maps as tools and recognisable points from which to launch their playful and sometimes intense explorations. In recent years, London artists have found inspiration in the map of the city; an inspiration that underpins this book. Disquiet with too faithful a representation has led to a delicate unpicking of the usually taut relationship between map and physical or social reality, and their explorations have uncovered many more geographies lying within the familiar terrain of the city.

LONDON: CHANGE AND GROWTH

MAPPING THE CITY

Maps are spatial images, but they have a temporal quality as well. They fix a moment in time, not always accurately, but a precise moment nonetheless; a quality that they share with photography but few other representational forms. This is clear on the earliest maps we have of London, such as the anonymous Copperplate Map, circa 1550—which depicts individuals active on the outskirts of the city, laying out washing, practising archery and carrying goods to and from market—but is just as true for later maps, up to and including the most recent interactive navigational systems that attempt to provide up-to-the-minute information on road works and traffic jams for harassed drivers.

This temporal specificity is invaluable to historians who use maps as evidence and illustration of physical change and development as well as—conversely—to date maps using evidence available on the ground or from other records. Inaccuracies and consistent errors, which run against the historical record, also leave trails of evidence connecting one mapmaker to another and reveal a general tendency to copy rather than to survey afresh with each map. Mistakes (and sometimes political forgetfulness) abound on maps, sometimes necessarily, but occasionally through lack of care or effort—but nothing changes their extraordinary relationship to the moment that they portray.

Maps act as markers that gradually reveal a narrative as to how a place has developed. Each map reports from a different temporal perspective, and series of maps encourage exploration and an individual understanding of both place and event, enabling the explorer to cut through the varied, and not always pure, motivations of the mapmakers. Maps, with their multiple over-layered levels of information, provide a parallel to the intricate interweaving of narratives that make up the story of London, and it is often the concision required to encapsulate all relevant information on a single sheet that makes them so fascinating and telling.

Cartographic markers arrive late in the history of London, springing into almost fully formed life in the reign of Elizabeth I, 1533–1603, when the Copperplate Map, of which plates for only three sheets out of a likely 20 survive, appeared. It was followed by a smaller scale map, created by Braun and Hogenberg, which was apparently copied from the Copperplate in 1572, and later, Francesco Valegio's woodcut of circa 1580 and John Norden's of 1593, as well as other revised versions deriving from these.

All holyes ni the Voall.

BVSSHOPPES GATE

MOOR FIELD.

Dogge hows.

Giardin di Pietro

Bedlame

Bedlame Gate

St Bintoth

Blak hows.

S. M.ª Spittel

SHORDICHE

Bussfioppes gate strete;

Copperplate Map (Moorfields), detail
circa 1556–1558

This map, originally believed to have been on 20 sheets, is now only known through three engraved copper plates that survived as the backing for oil paintings. This part of the map has an image of the Tower of Babel attributed to Martin Van Valkenborgh, 1535–1612, on the reverse. Likely to have been produced in Antwerp, it is the first known printed map of London and was also the source for both the Hogenberg and Agas maps.

Buildings are shown in semi-perspective, with a mixture of individual attention for the more significant, and stylised repetition for the less so. Clearly the mapmakers had spent a considerable amount of time visiting and sketching the more publicly accessible places including, on this sheet, the lunatic hospital of Bedlam (before its destruction by the Great Fire in 1666), all Hallows in the Wall and the already dissolved hospital attached to St Mary Spital, before they returned to the Low Countries to prepare the sheets for engraving.

On this map, the City is still enclosed within a strong defensive wall and crowded with houses, churches and commercial buildings. There are gardens belonging to the wealthy, or religious establishments, but they are necessarily enclosed within their own walls. In contrast to the extramural areas, the city streets are deserted.

Image courtesy of the Museum of London.

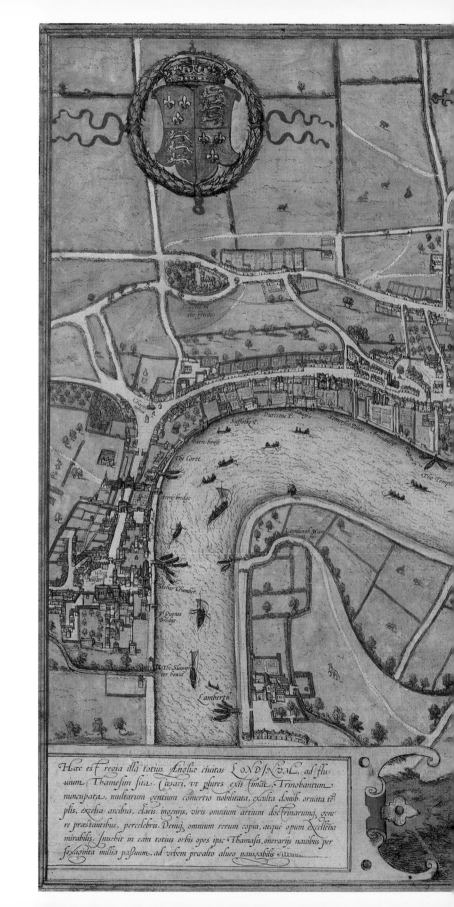

Londinum Feracissimi Angliae
Regni Metropolis, 1572
Georg Braun and Frans Hogenberg

This is one of the best-known maps of Tudor London, which shows the city just before Elizabeth I took the throne in 1558. Its close similarity to the parts of the Copperplate Map that have survived, means that it is highly likely that this was a single sheet version of the larger and more exclusive original. It was first published in a German atlas of European cities, the *Civitas Orbis Terrarum* in 1572. The colour, as with all maps produced until the advent of lithography in the nineteenth century, is hand applied and may have been added much later.

The image represents a true plan, but it has been disguised by the three-dimensional rendering of the buildings and the boats on the Thames and by the notion that it is a panoramic view from the hill that the four figures in the foreground parade on. The idea of the plan was clearly familiar to the original surveyors but it was not one that was acceptable to a popular audience.

This map, as with its predecessors, shows London bursting out of the confines of the city walls—despite the sudden availability of land within them—arising from Henry VIII's dissolution of the monasteries. The City controlled some of the Wards Without, including Portsoken to the west and notionally Southwark over the river, but it neither could nor wanted to resist the force of development. In addition, it didn't manage to achieve administrative control over Southwark, which was soon to establish a long-standing reputation as dissolute, lawless and louche. Already evident on the south of the river are the bull and bear baiting rings that would soon be joined by the Globe and Rose theatres. It is this connection with Shakespeare, together with its ready availability, that has made this map of London the familiar image that it has become.

Image courtesy of Ashley Baynton-Williams.

Opposite and right:

London and Wesminster, 1593

John Norden

Norden published this pair of images of the Cities of London and Westminster, along with another of the whole of Middlesex, in 1593, in *Speculum Britanniae*, a grand and ambitious plan to map the whole of the country that failed, at least partly, for lack of finance. Both maps are lively, if somewhat cartoon-like, and show individual buildings in great detail. Norden seems most preoccupied with recording the great houses and palaces that lined the Thames at this date. The owners of such were presumably the target of his fundraising efforts, which may also explain why he decorated the London plan with the Coats of Arms of the 12 prominent livery companies. Norden was neither the first nor last to exploit maps for their potential to flatter. These maps also include keys to various sights and important buildings, an innovation that was to prove one of cartography's most profitable and enduring.

The equal emphasis on both Cities was also an innovation. The built up area that now connected the two had created enough development around the Abbey and Palace complex at Westminster to make mapmaking a worthwhile activity.

Opposite: Image courtesy of Ashley Baynton-Williams.

Right: Image courtesy of David Hale/MAPCO.

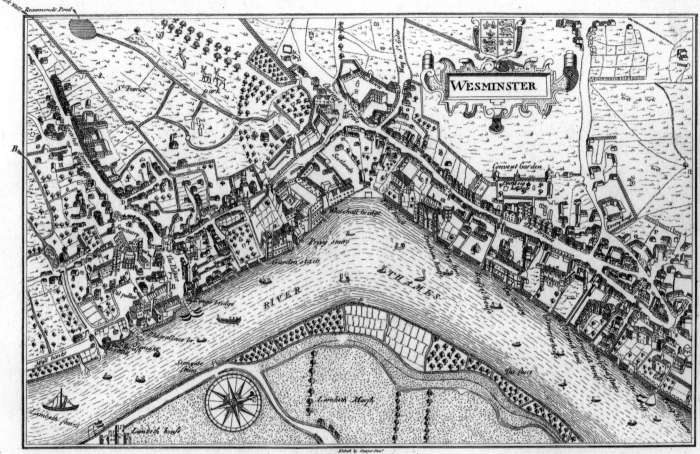

PLAN OF WESTMINSTER; FROM NORDEN'S SURVEY, TAKEN IN QUEEN ELIZABETH'S REIGN, 1593.

Herein are comprehended all the buildings from Temple Bar to Mill-bank Street, when an uninterrupted view of Primrose Hill, Hampstead, Highgate, &c, might be had from the houses on the north side of the Strand. By the several water-courses which issue from Rosamonds Pond and elsewhere, and from the one at B, which runs through the present College Street, and under Mill bridge to the Queen's Slaughter House, it appears probable that Thorney Island was more extensive than is supposed; but of this, more elsewhere. N.B. A Standing Gibbet is seen in this plan, in New Palace Yard.

A The Abbey
B Westminster hall
D Long ditche
E Thorny Lane

F The Amneris
G The way to Tothill Fielde
H The Lord Dacres
K Kings Streete

London, Published as the Act directs February 20th 1808, by John Thomas Smith, No.31, Castle Street, East, Oxford Street ——

L Round Woalstaple
M The Parke lodginge
N The Tilt Yard
O S Martynes in the field

P Clements Inne
Q New Inne
R S Clements Danes
S Temple barre

That we only know of the Copperplate Map, because the extant plates were reused as the base for otherwise unrelated paintings, encourages the thought that many others may have also existed and completely disappeared. However, because it was widely copied by Braun and Honenberg and others, and there appear to be no competitive sources, there is the suggestion that it may have been unique. If so, how did pre-Elizabethan Londoners, and the many visitors to the city from across the world, understand London as a place, and why weren't they tempted to draw enough sketch maps of the streets and buildings so that at least one might have survived?

Enough such maps and plans have endured the previous three centuries, often depicting small towns and estates outside of London—indicating that the drawing of plans for practical purposes was not unknown or simply confined to the metropolitan elite. But still, the few early images that exist of London are either perspective views such as those presented by Wyngaerde, circa 1544, or by the tiny thumbnails on coins and medals, the clusters of buildings in the background of illustrated manuscripts and, perhaps more significantly for this subject, as symbols on travel, pilgrimage and religious maps.

By the sixteenth century, maps of the world were, in contrast to the more immediate urban realm, fairly fully developed. Sailors possessed extensive maps showing how to navigate the world, with all the necessary information on ports and coastlines. Pilgrims kept informative route maps derived from Roman examples that helped them, principally, to get to and from Jerusalem and, later, Rome. The requirements of national defence and governance had also begun to generate maps, often resulting in a concentration on ports and the coastline, including—critically—estuaries and particularly that of the Thames. Travellers were a good deal better informed on how to cross the globe to reach London, rather than in how to locate their final destination, once they had arrived. Perhaps common sense, a good memory and asking directions was considered adequate. Certainly, the instructions given to Sebastian by Antonio in Shakespeare's *Twelfth Night* (notionally about Illyria but clearly referring directly to London) would indicate that navigation had to be achieved with minimal information and without help from even the sketchiest of maps: "In the south suburbs, at the Elephant, is best to lodge: I will bespeak our diet,

Civitas Londinum (Agas' Woodcut Plan)—A Survey of the Cities of London and Westminster, the Borough of Southwark and parts adjacent in the Reign of Queen Elizabeth, 1633

Civitas Londinum is another version of the Copperplate Map, freely interpreted and printed from woodcut blocks on eight sheets. It is known from three copies, including one in London's Guildhall. Its attribution to the Sussex land surveyor, Ralph Agas, has long been known to be erroneous, but the label has stuck and it is still commonly known as the Agas Plan. The care for detail in the Copperplate Map is not evident here and, in many ways, it is a cheap and cheerful—if invaluable—view of the city showing landmark structures amongst a mass of crudely drawn houses and other non-specific buildings.

The map extends from St James Park in the west to Aldgate in the east. To the north, the view stretches away in perspective up hills ringing the London basin. To the south, the map barely crosses the river. But because of its derivation from the Copperplate Map, its comprehensive quality and status as the only large-scale map of sixteenth century London, *Civitas Londinum* is the essential view into the teeming streets of Tudor London. It shows us the interrelationship between churches and streets and houses and how, outside the walls, London rapidly transforms into fields and villages. It shows the windmills that must have dominated the horizon and the regular lime kilns that filled the sky with smoke. For all its naivety, it provides a recognisable and vivid view—albeit from Restoration London—into a past which was beyond living memory.

Image courtesy of the Guildhall Library, City of London.

while you beguile the time and feed your knowledge with viewing of the town: there shall you have me."

The breakthrough in the production of maps, which drove their sudden proliferation, was the arrival of printing. Imported maps were available from the 1530s, including the Copperplate Map, which was probably engraved and printed in Antwerp. The availability and popularity of relatively inexpensive maps triggered a new market in sales and production, resulting in the first maps being printed in England during the late sixteenth century. These new maps include the large-scale woodblock copy of the Copperplate Map ascribed, although without much evidence, to the surveyor Ralph Agas, which was probably printed in 1633.

The earliest maps of London show the compact city enclosed within its walls with various gates, most still familiar as place names in contemporary London, providing access to the roads and fields beyond. Development can be seen working its way along the main roads out of the city and across the only bridge over the Thames into Southwark. The City of Westminster is still a separate town, two miles upstream, but is firmly linked to the City by continuous development along the Strand and Fleet Street.

In 1560, the population of London was estimated at approximately 120,000, although figures are far from accurate or easy to establish. Despite the low life expectancy of Londoners, and the regular outbreaks of plague throughout the sixteenth century, the population had more than doubled since 1500, from 40,000–50,000, and would triple again by 1650, to 370,000, just before the last great outbreak of the plague in 1665. By 1550, London was set on a pattern of growth that would transform it from being a town significantly smaller than Paris, Antwerp and many Italian cities, to being the largest, and arguably the greatest, city in the Western world. In the period 1500–1650 it achieved a ten-fold population increase—easily enough to put a near impossible strain on its systems of governance, trade, supplies and waste disposal.

The coincidence of this extraordinary growth, with the sudden availability of land both in and around the city—resulting from Henry VIII's dissolution of the monasteries in 1539—meant that the supply of land for new development was not an immediate problem (although it did simultaneously lead to the disappearance of several of London's hospitals and much of its health care provision, which then had to be expensively re-established and rebuilt by the city fathers). The new supply of land led directly to the denser city development shown on maps from the beginning of the seventeenth century onward. The estates of 23 religious orders, within or immediately around the City walls, were largely sold off to fund the various wars initiated by Henry VIII. The transfer of so much land and property in the city to the open market, led to far greater freedom to develop and change, and a much more intensive use of the land. It permitted new housing and commercial uses in London streets and the removal of religious control also meant that theatres, and other places of entertainment, which had previously been restricted to the south bank of the river, could take their place in the streets of the city proper for the first time.

The moment of the restoration of the Monarchy in 1660, marks a dramatic shift in mapmaking in London. This was due both to the fervent intellectual climate in the city (when Newton, Wren, Hooke, Pepys, Evelyn and many others found themselves in a close-knit circle prepared and able to take on any subject) and increase in demand. Suddenly, accuracy was critical and a new style of presentation, instantly recognisable to us today, was developed and became the established norm.

Richard Newcourt had produced a new large map of London, in 1658, during the Commonwealth period but it still adhered to the old style—the bird's-eye view of houses and streets. Lacking accuracy, it also unimaginatively repeated the same house form—apparently *ad infinitum*—across London, giving it an entirely different character from the one it was intended to portray. Wenceslaus Hollar, one of London's greatest topographical artists, had also begun a survey and was using the same bird's-eye viewpoint, although he was drawing, in great detail, real buildings in place of Newcourt's cartoons. Hoping to secure subscribers and sponsors, he had already engraved a preliminary plate in 1658 showing west and central London, but had got no further with the project by the time Charles II returned from exile.

Within a few years of the Restoration, change was happening on many fronts. In 1662, Samuel Pepys—having successfully made the tricky transition from republican into monarchist in his new role as Clerk of the Acts to the Navy Board—was engaged in commissioning the production, by Jonas Moore, of a new chart showing the Thames from the sea up to Westminster and beyond. In the manner of naval charts, this was an accurate two-dimensional planar representation

AN EXACT SVRVEIGH OF THE STREETS LANES AND CHVRCHES CONTAINED WITHIN THE RVINES OF THE CITY OF LONDON FIRST DESCRIBED IN SIX PLATS BY IOHN LEAKE, IOH HENNINGS, WILLIAM MARR, WILL. LEYBVRN, THOMAS STREETE, & RICHARD SHORTGRAVE in Dec.ʳ A.ᵒ 1666. BY THE ORDER OF THE LORD MAYOR ALDERMEN AND COMMON COVNCELL OF THE SAID CITY.

Reduced here into one intire plat, by Iohn Leake. the Citty Wall being added al placet where the Halls stood, are expreſt by Coats of Armes, & all the Wards, divided by pricks & Alpʰ

The Prospect of this Citty, as it appeared from the oppoſite Southwarke ſide, in the fire time.

The Right Honourable Sʳ William Turner the Lord Major. A.ᵒ 1669.

Artillerie ground.

Moore Fields

Moore gate.

Hatton Garden

South Field

Guild Hall

Aldersgate

Chapeside

Paules Churchyard

Ludgate

Bridewell

Tower hill

East Smith Field

THE RIVER THAMES

Part of Southwarke

a Scale of Feet.

Published with the description of the Warde, by the care Industrie and Charge of Nathanaell Brooke Stationer, and are to be Sould

A Map or Grovndplot of the City of
London within the Suburbes
thereof, 1666
Wenceslaus Hollar

In 1666, Wenceslaus Hollar had already been drawing and engraving maps and views of London for several years. An immigrant from Bohemia, he had passed through Antwerp en route to position under the patronage of the Earl of Arundel—bringing considerable topographic and artistic skills with him. He had developed the highly accurate axonometric plan views of London and published the first in a series of detailed area studies in his engraving of west central London which was eventually intended to cover the whole of the city. He also surveyed and began to engrave the city in the same style when the Great Fire struck.

The Fire was Hollar's grand opportunity. He produced several views of it from across the river, but more significantly, within a matter of weeks, he had also prepared a 'groundplot' of the burnt and devastated area. The groundplot was a new departure in presentational terms and marked the introduction of a simple map to the general public. The surrounding streets are shown from the birds-eye view typical of Hollar.

Hollar went on to become a part of the official survey team—lead by John Leake—that mapped the burnt area for ownership, insurance and rebuilding purposes and he produced similar plans of greater and greater detail throughout his lifetime.

of the river, and the streets and buildings of London are drawn in the same way, as a ground plan. Although it seems likely that Moore would have copied the plan from elsewhere, it is the first accurate plan of the city we have and is the style that became the commonly accepted and almost exclusive way of presenting a map. Only recently have mapmakers returned to bird's-eye view representations.

Jonas Moore had learnt the surveyor's trade mapping and draining the fens from 1649 onwards. In 1660, he returned to London and, in 1663, joined the group of scientists, mathematicians and experimenters who were founding the Royal Society. Such men were all engaged in the art of observation and precise graphic description of what they observed. In 1661, Robert Hooke and Christopher Wren were amusing the new King with accurate drawings of a flea, a louse and the wing of a fly, as observed under a microscope and later published in Hooke's *Micrographia*, 1665. The stars were being observed and mapped by Halley and others, and the problem of measuring longitude was fuelling research into all sorts of new technology and equipment. It was an age of study using the latest scientific instrumentation and the meticulous recording of the results.

All of this spilled over into the study and mapping of London; an activity involving many of the same individuals.

However, Moore's naval map, and the influence of the Royal Society, may have not been enough to change the way mapmakers drew cities; this required a more potent trigger. By 7 September 1666, the Great Fire of London, which took place over five days, had savaged 373 acres of the 448 acres within the walls of the City, as well as a further 63 acres westwards of the old fortifications. The fire had even managed to cross the Fleet River before coming to a halt close to Fetter Lane. Wenceslaus Hollar—his survey of the destroyed city already complete, and the surrounding buildings drawn in the old style—rapidly produced an accurate ground plan of the burnt-out area, which he referred to as a "groundplot", marking roads and "the churches and other remarkable places". On 22 November 1666, just 11 weeks after the Great Fire, Pepys was shown this new plan and wrote in his diary:

My Lord Brouncker did show me Hollar's new print of the City, with a pretty representation of that part which is burnt, very fine indeed; and tells me that he was yesterday sworn the

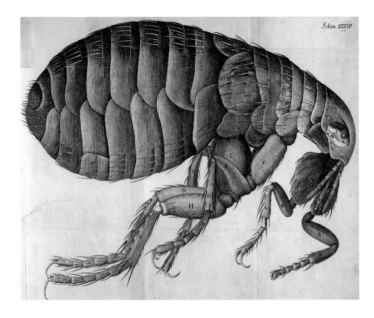

A Flea, illustration from
Micrographia, 1665
Robert Hooke

From 1664–1665, Robert Hooke and Christopher
Wren were using the newly available microscope to
examine and record minutiae as accurately as they
were able. The fledgling Royal Society published
Hooke's observations in *Micrographia*. Both Wren and
Hooke, together with other members of the Society,
brought the same rigour to their studies and maps of
London when it came to producing plans following
the Great Fire in 1666.

King's servant, and that the King hath commanded him
to go on with his great map of the City, which he was
upon before the City was burned, like Gombaut of Paris,
which I am glad of.[1]

Hollar's map was immediately recognised as 'news', at least by the
Dutch with whom the English were at war, and was reproduced by
de Wit in Amsterdam for distribution across Europe. Hollar later
became part of the team of surveyors—appointed by Charles II under
the mathematician John Leake—to fully survey the burnt out area
and properties. The result was a large map published in 1669, again
engraved by Hollar in his familiar style. This further map also had
a new agenda—administration and land tenure—and accuracy was
critical. No longer were such maps chiefly decorative or produced
for connoisseurs and collectors; the main purpose was now ruthlessly
functional and intended to facilitate the rapid rebuilding of the city.

But before Hollar had even engraved his first plan or any of his
various views of the Fire in full destructive force, others got there first,
employing the ground plan with even greater purpose. Christopher

Wren and John Evelyn each produced new designs for the destroyed
area and presented them to the King on 11 and 13 of September,
respectively. Later that year, Hooke, Newcourt and Valentine Knight
all produced schemes. The inspiration in this instance was probably
the architectural plan, but the effect was the same, a ground plan that
was the new orthodoxy and would stay so for many centuries to come.

Hooke, energetic and restless, was heavily involved in the
huge proliferation in map production that followed the Great Fire.
With many stocks of prints and plates destroyed in the shops around
St Pauls, there was an urgent need to produce new material. The
publisher John Ogilby, who had lost almost his entire stock of books in
the Great Fire, went into partnership with his step-grandson, William
Morgan, as both producers and publishers of maps. *Ogilby's Book of
Roads* was published in 1675 and followed by A Large and Accurate
Map of the City of London, 1676, which was produced at a scale
of 30.5 metres to 2.5 centimetres and printed on 20 separate sheets
(plus a sheet with a dedication to Charles II) shortly after Ogilby's
death at the age of 76. Hooke and Ogilby held almost daily meetings
in the course of the map's survey and production, the result being the

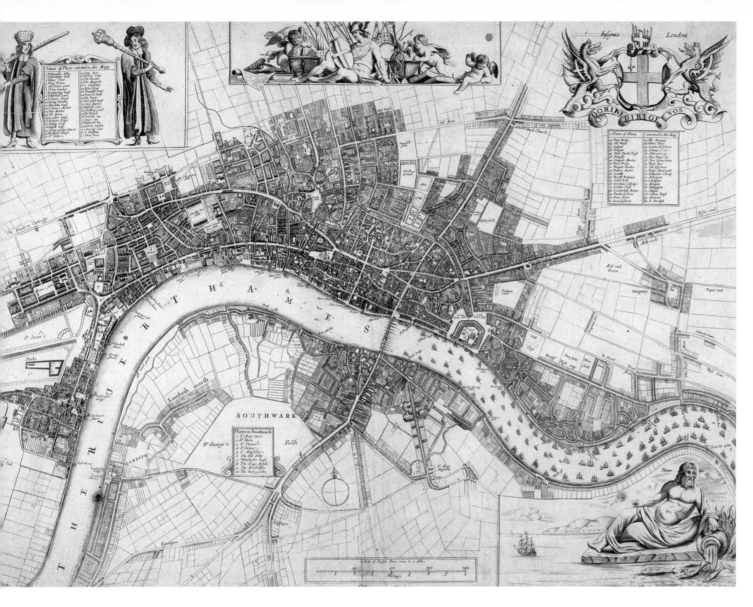

London and Westminster (London Rebuilt)
—A Mapp of the Cityes of London and
Westminster and Burrough of Southwark with
their Suburbs as it is now rebuilt since the late
dreadfull Fire, 1680
John Oliver

Despite the visionary plans put forward by members of the Royal Society (and others) for the redevelopment of the city after the Great Fire, the need to rebuild quickly meant that existing roads, boundaries and site ownerships were respected as far as possible and the results were not so different from what had existed before. Oliver's map of 1680 demonstrates how soon business was back to normal and how the relentless consolidation and expansion in London continued. City buildings are squeezed in between narrow and twisting Medieval lanes while the city pushed west with a new, markedly orthogonal and rationally planned development along Pall Mall and up into St James.

Image courtesy of the Guildhall Library, City of London.

first entirely accurate map of London. It shows in detail the ground plan of the city before the Great Fire with each property and open space carefully delineated. It remained the *de facto* source map of the city for another 70 years, supplemented by William Morgan's further surveys of Southwark and Westminster.

By the end of the seventeenth century, the population of London was approximately 575,000. North of the river, the City and Westminster are linked by solid and compact development, with a very tightly drawn boundary where the dense streets stop, and the edge of the town meets pasture, parkland and heath. The first London squares have appeared: Lincoln's Inn Fields, Covent Garden Piazza and St James' Square amongst them. St James', Hide (sic) and Marylebone (later Regent's) Parks are all in evidence. To the south, Southwark—still connected to the City by only the one bridge—has grown with development continuing along all roads out of town. The marshy land between Southwark and Lambeth is just beginning to be drained, laid out and built on. To the east, the city growth extends along both banks of the Thames, emphasising the importance of both national and international trade activity to London's growth and success.

By the mid-eighteenth century London becomes two-fold again. Although the Cities of London and Westminster are now joined by solid development, there is a critical difference between the elegant Georgian sector growing in the west and the poverty and tumult of the east. The two worlds, despite intersecting and overlapping with one another, are different and distinct. They are the worlds crossed into by Hogarth's protagonists in his *Marriage A-la-Mode*, *Rake's Progress* and *Harlot's Progress*. Affluent Georgian London is a brilliant and decadent place populated by aristocratic society, wits, inventors and some seriously rich bankers and merchants. Poor London is gin-sodden and where Hobbesian short, brutish lives are lived out in filth and squalor. The new terraced squares of the West End co-exist with the decaying and tottering slum districts.

The city was a magnet for all sorts of people. The gentry (also known as the 'Quality') felt obliged to be in London for the season and own, or at least rent, a property in one of the fashionable areas. The Scottish were arriving in great numbers, as did many others from across England. Irish immigrants became the predominant residents of the St Giles rookery; there was a large Jewish community growing

Following pages:

Plan of the Cities of London and Westminster
and Borough of Southwark, with the Contiguous
Buildings, 1749
John Rocque

After a dearth of new surveys and maps of London covering 70 years, mapmaking springs back to vigorous and scintillating life with the publication of John Rocque's magnificent 24 sheet map showing the early Georgian city in confident and expansive mood.

The survey work, undertaken with the engraver John Pine, took seven years in total and must have caused intense financial stress to Rocque during this period. None of this is apparent however, on the maps themselves, which are carried out with the highest possible degree of accuracy and attempt to clean up the seedy image of London as presented in such novels as Henry Fielding's *Tom Jones*, 1749, or in William Hogarth's paintings and popular engravings of the same period. Shown here is the slightly later 1749 edition of Rocque's map, a result of the continuing popular demand for reproductions of the new map.

Image courtesy of Ashley Baynton-Williams.

in Whitechapel; and the Huguenot refugees, who had escaped persecution in Louis XIV's France, were settling in Spitalfields.

John Rocque, London's greatest mapmaker of the period, and one of a family of Huguenots who arrived in the first decade of the century, trained as a *dessinateur de jardins*. He was employed by landowners to measure and draw their estates, including landscapes by the designers Capability Brown and William Kent. When he came to survey and publish his magnificent maps of London he translated the idea of the figure-ground depiction of the streets into the eighteenth century landscape ideal. Rocque's depiction of London—on 24 sheets—is a picture of elegance and clarity, a tapestry and delightful pattern of geometry and picturesque incident. The texture of the many parks, gardens and fields is rendered with almost fetishistic obsession, and the dense pattern of streets and blocks of the city could as easily be planter beds as crowded tenements. Rocque's maps reflect the view of his middle and upper class contemporaries—that their London was 'the new Rome'—and incidentally bear a resemblance to the famous maps of 'old Rome' by his near contemporary Giambattista Nolli.[2] In such maps, there was no room for the poor, the danger, sickness or grime that is the focus of much contemporary writing about London.

Rocque's map, despite its sanitised view of the streets, was certainly accurate and based on the first new survey of London since the 1670s. When it was first published in 1746, Ogilby and Morgan's map of 1676 had been recycled, copied and re-issued innumerable times by less energetic mapmakers, often with the cartouche announcing that it was "new and exact". Rocque rapidly followed his 24 sheet masterpiece with a similarly astonishing map of London encompassing "the adjacent Country 10 miles round". This map gave him even more opportunity to depict the landscape surrounding the city's densely built centre, and to provide a clear outline of its continual growth and absorbtion of outlying towns and settlements into city suburbs.

Rocque's maps reveal a city obsessed with its own image. By 1750, with a population of over 675,000, London was easily the most crowded city in Europe and had begun to overtake other cities such as Cairo and Constantinople in this respect. It was certainly the wealthiest city anywhere. Yet it still had no obvious centre, nor any significant streets or civic spaces to express as such.

Developments were largely of a practical rather than triumphal nature. A bridge was being constructed across the Thames at Westminster (to be opened in 1750)—the first new bridge since the Romans built London Bridge—and, in 1756, the world's first bypass 'The New Road' (now the Marylebone and Euston Roads) would be cut along London's northern boundary to relieve traffic congestion in the centre.

The city, however, was expanding and—despite the construction of elegant squares and terraces to the west, and building taking place on the land to the east, including Spitalfields—it was still a city that, in 1763, the *London Chronicle* claimed could be walked around in seven hours. Most of the additional population had found homes in the increasingly dense streets of the existing metropolis and open space, probably to Rocque's chagrin, and was being sacrificed to more houses and commercial buildings. The perimeters of town still weren't far away and it was there that Londoners went in their moments of leisure, whether to spas or entertainment centres such as the Vauxhall and Ranelagh 'Pleasure Gardens', or simply to open parks and fields.

From the turn of the century and the publication of Richard Horwood's map, 1792–1799, the publication of new maps and changes to the city struggled to keep up with one another. New maps, often based on previous designs but augmented with new detail, appeared at a rate of at least two a year. To make this possible, they aimed at different potential markets; travellers, tourists, landowners, walkers and those who wanted to calculate the cheapest road tolls. London was finding new ways to grow and development had crossed former boundaries such as the New Road or into the drained marshland south of the river. This was aided and abetted by more effective modes of transport and newly opened roads, such as Regent's Street, cut straight through the centre of London. Multiple bridges now crossed the Thames, while stagecoaches and Hackney cabs operated a relatively efficient service across the city. The next revolution was expected imminently with the arrival of the railways. Tracks were already being cut through the outlying suburbs.

By 1827, maps also communicate the huge changes taking place in commercial and industrial London. Canals connecting London and the Thames with the industrial North were running across north and east London. Large docks cut into the land, on either side of the river (including the London, East and West India and Surrey Docks), in response to ever-expanding trade and the congestion crisis of the river and wharfs in the 1790s.

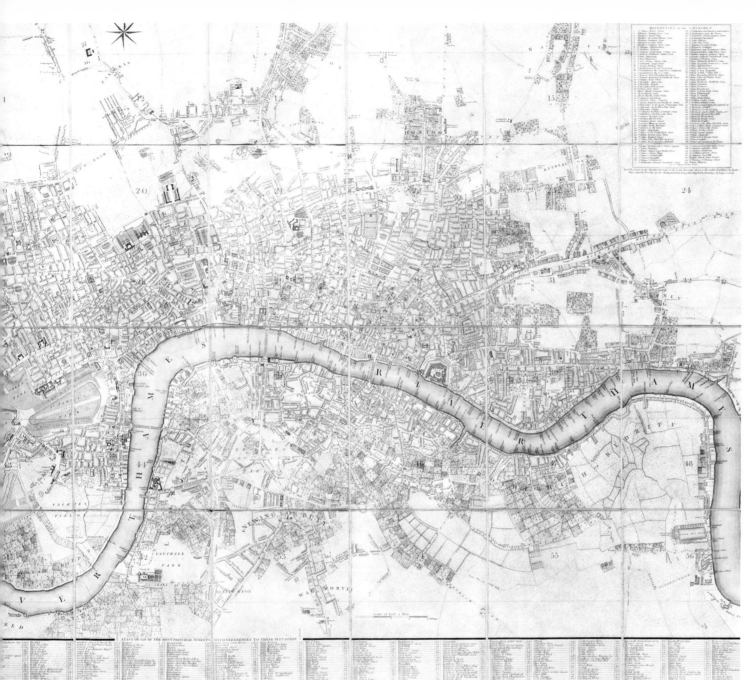

Cary's New and accurate Plan of London, Westminster, the Borough of Southwark and Parts adjacent, 1787 (1799 edition)
John Cary

Folding, linen-backed maps of London begin to appear in great numbers during the second half of the eighteenth century and this map, published by John Cary, is a good example of the period. Cary was a prolific map producer and one who carefully updated his maps for each new edition. This large map went through multiple editions following its original publication in 1787, with a final 'corrected' version published in 1825. The hand colouring was undertaken at the time of issue and the map was sold to the public in a standard slip case. This is a very different product to the printed maps of only a few years earlier, such as those intended for wall-mounted display or table-top consultation—it is a map designed for real navigational purposes.

Cary's map is not concerned with extraneous detail—the fields beyond are barely filled in except for some useful tracks and by-ways. St James' and Hyde Parks are, however, an exception—not only as a result of the applied colour. They are lavished with attention, thus becoming the most important attraction in town—befitting their role in the life of London at this time.

Image courtesy of Ashley Baynton-Williams.

Plan of the Cities of London and Westminster, the Borough of Southwark and Parts adjoining shewing every House, 1792
Richard Horwood

The large map of London, produced by Richard Horwood in 1792, was nothing if not ambitious. Horwood wanted to show every house, including its house number, in London. The former he achieved, including all the courts, alleys and gaps in development—marking a breakthough in accuracy and detail—but the latter proved too onerous and was never completed, even in later editions.

Horwood was working for the Phoenix Insurance Company on occasional surveying jobs when he started work on the map, and he clearly perceived its relevance to the many insurance companies, parish vestries, commissioners of sewers and others whose work it was to keep records, administer works and provide services in the city. Most of these, up to and including King George III, subscribed to the map.

The map's detail and the identification of so many individual properties has led to it being described as the "most important London map of the eighteenth century" (Howgego, 1964) and it is a vital aid to historians of the period. Horwood never re-issued the map and died in poverty in 1803 at the age of 45. William Faden published further editions in 1813 and 1819, extending its coverage to show the new works to the docks in the east. It remained the most accurate map of London for a further 50 years until the publication of the first edition of the Ordnance Survey in the 1860s.

Image courtesy of the Guildhall Library, City of London.

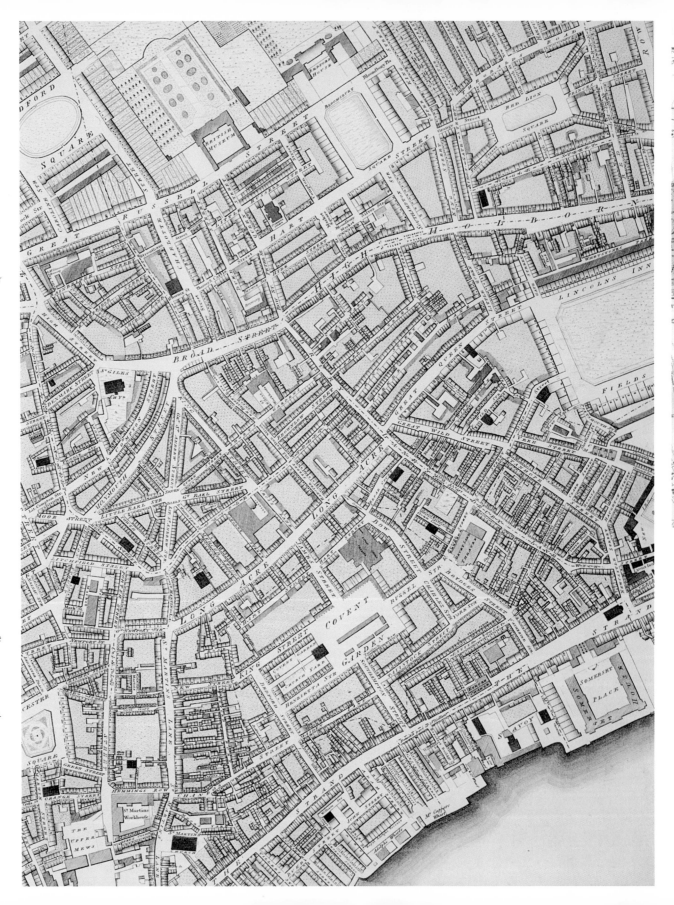

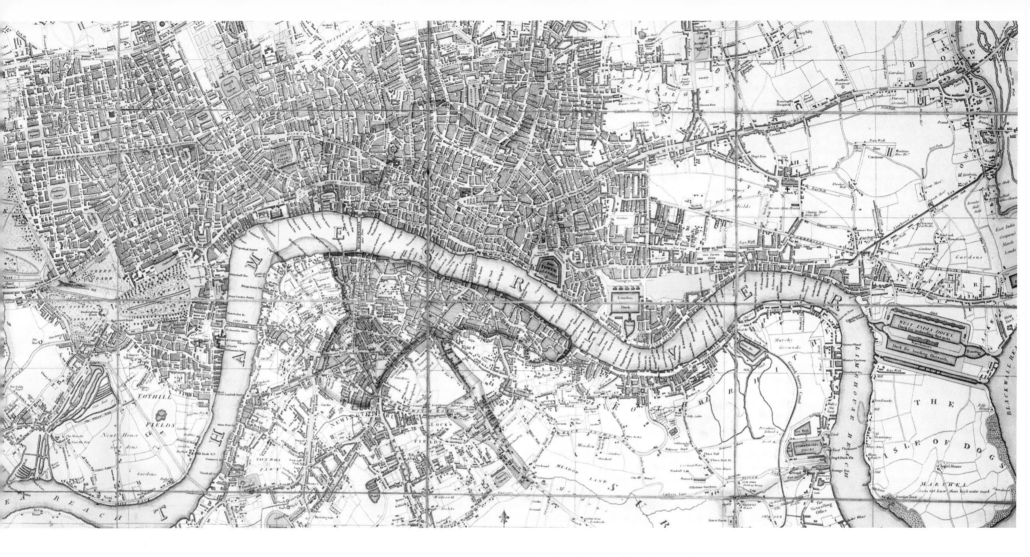

A New Map of London, Westminster and
Southwark, 1804
Robert Laurie and James Whittle

This is one of a myriad of London maps issued at the
start of the nineteenth century. Laurie and Whittle
published their first London map in 1776 (having taken
over the plates of a 1765 map by Robert Sayer), but
public interest was now focused on east London—
particularly the creation of the city docks—and
mapmakers were required to meet such demands, by
extending their maps at least as far as the Isle of Dogs.

Laurie and Whittle's map shows the twin West India
Docks cutting across the narrow top of the Isle of
Dogs—only just completed in 1802. Later maps would
show further dock basins in the area and year by year
infrastructure changes in London would become the
driving force for mapmakers.

Image courtesy of Ashley Baynton-Williams.

LANGLEY's FAITHFUL GUIDE through LONDON and places adjacent, in every direction from St PAULS

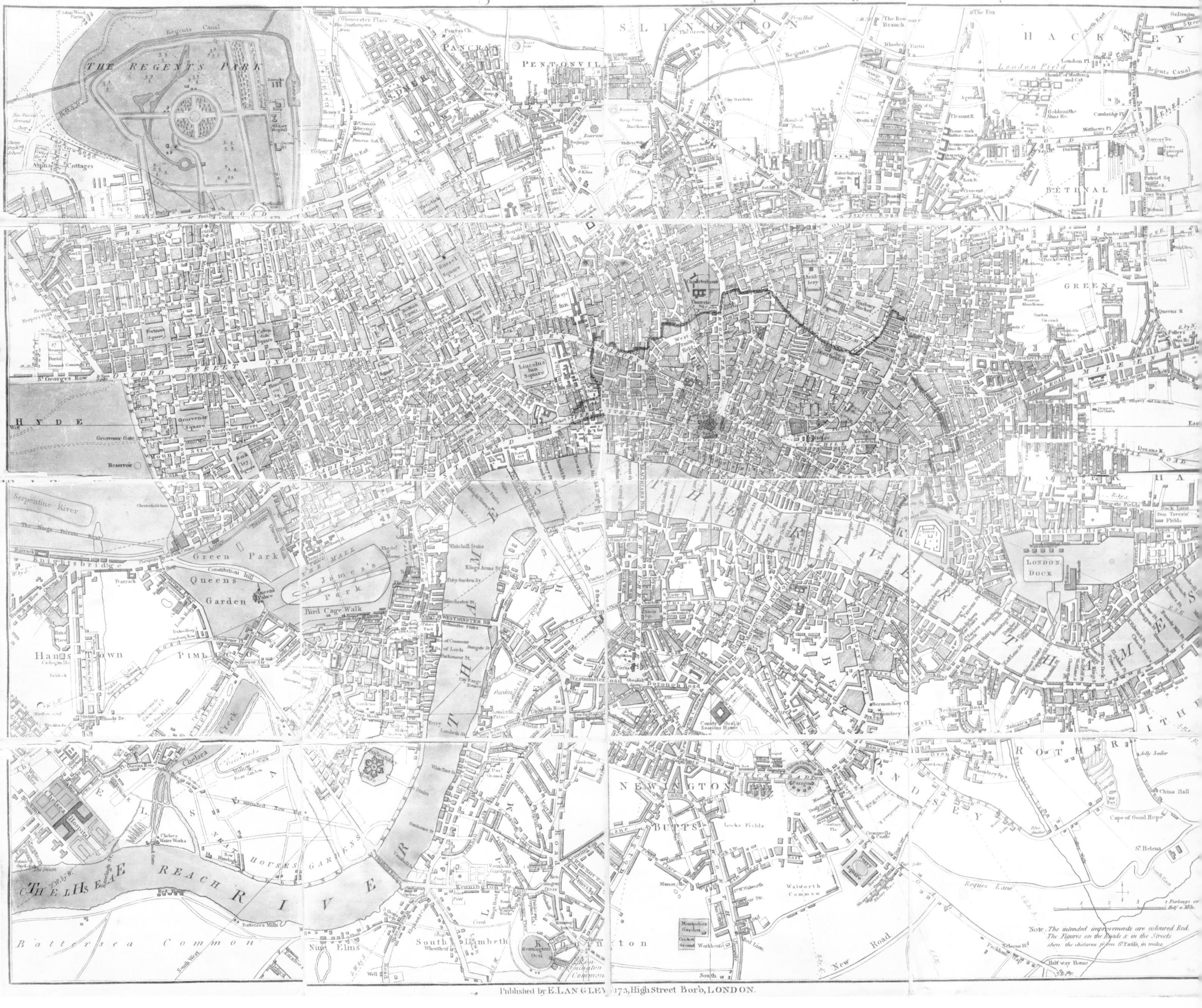

THE REGENT'S PARK

HYDE

Serpentine River

Green Park

Queens Garden

St James's Park

Bird Cage Walk

WESTMINSTER BR.

CHELSEA REACH

RIVER

Battersea Common

Nine Elms

South Lambeth

Kennington Oval

Lincolns Inn Square

PENTONVILLE

HACKNEY

BETHNAL

GREEN

LONDON DOCK

ROTHERHITHE

NEWINGTON

BUTTS

Note. The intended improvements are coloured Red.
The Figures on the Roads & in the Streets
shew the distance from St Pauls, in miles.

Published by E. LANGLEY, 173, High Street Boro, LONDON.

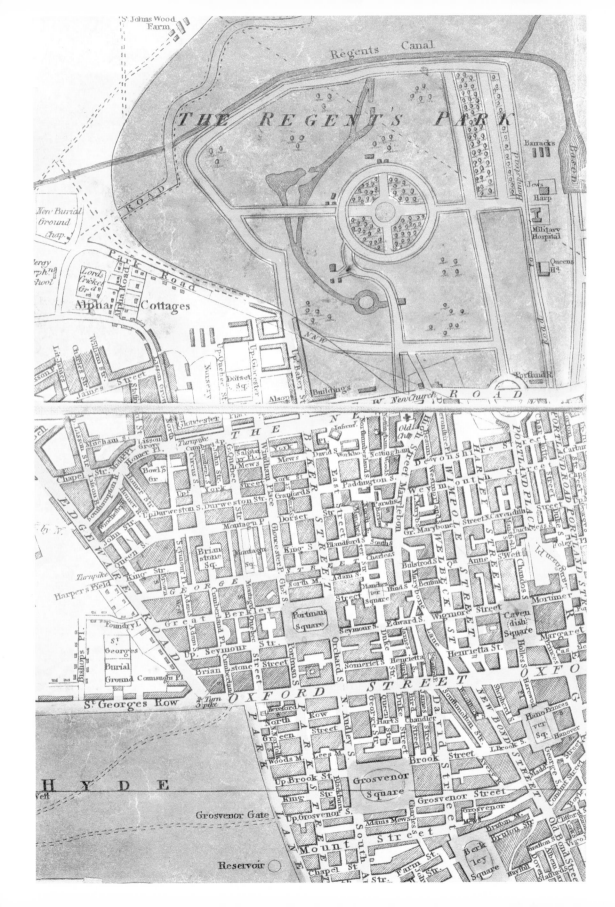

Langley's Map of London/Langley's Faithful Guide through London and Places adjacent in every direction from St Paul's, circa 1824
John Luffman and Edward Langley

Originally published by John Luffman in 1809, this map was taken over by Edward Langley following his split with his partner William Belch in 1820. Luffman, meanwhile, continued to publish other maps under his own name. (This is indicative of some of the competitive and commercial turmoil of the map publishing world in London at this time.)

Langley ensured the map was updated and it shows the Regent's Canal (completed in 1820) running through Regent's Park, Camden and Islington at the north of the map, and the series of new bridges over the Thames: Vauxhall (opened in 1816), Waterloo (opened in 1817)—labelled here as Stand Bridge, its pre-completion (and battle) name—and Southwark (1819) joining the existing London, Westminster and Blackfriars Bridges. The railway system can be seen approaching Euston and King's Cross Stations, although it would be many years before the stations themselves were built.

The main curiosity is Regent's Park itself, which appears to be the result of haphazard guesswork and completely lacks the distinctive shape it inherited from the layout of the original farm boundaries on the site. The park was in construction at this time according to John Nash's designs, and many of the surrounding crescents and terraces were already completed, so there seems little reason for Langley's eccentric approach.

Images courtesy of George Hulme.

CRUCHLEY'S NEW **PLAN** of **LONDON** IMPROVED TO 1827. INCLUDIN

REFERENCE
Boundary of the City of London
Westminster
Do Borough of Southwark

ENGRAVED & PUBLISHED BY G.F.CRUCHLEY. MAP SELLER.
(from Arrowsmiths) 38 Ludgate Street. London . 1827.

REFERENCE
Edge of the Kings Bench Prison.
Do Fleet Prison
Extent of the Rule Liberty

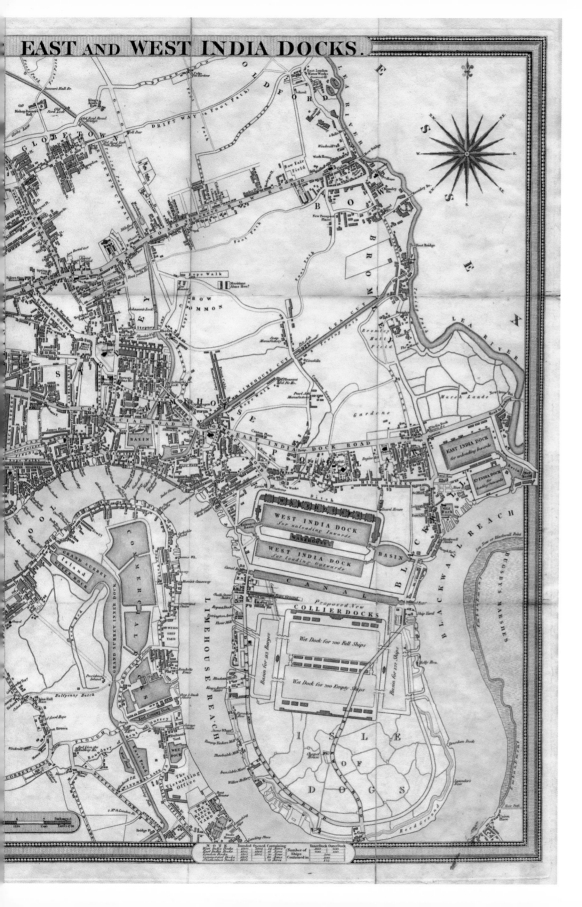

EAST AND WEST INDIA DOCKS.

Cruchley's New Plan of London improved to 1827 including the East and West India Docks, 1827
George Cruchley

This is one of the many maps of London available in 1827, and one of the best alternatives to the Greenwoods' map. This plan by Cruchley was first produced in 1826 and updated at regular intervals. Cruchley was eager to impress by including the docks—as ever at this period, a subject of huge public fascination. This map went on to be printed for the next 20 years, with dates and titles changing to suit.

Image courtesy of David Hale/MAPCO.

Large new factory buildings were beginning to appear, especially near the docks and canal basins.

This year, a year like any other in the period, saw new maps published by John Britton, John Bumpus, George Cruchley, William Darton, James Gardner, John and Christopher Greenwood and Samuel Leigh. Most of these went through many editions and, although the Greenwoods' map (on six sheets altogether) resulted in their bankruptcy, it went on to be published by E Ruff and Company up until as late as 1854. By then, it showed a London that we are very familiar with; railway lines cross the outer suburbs, which themselves are filling up with row upon row of small Victorian cottages. London's major new park, Regent's Park (in construction from 1812 on) by John Nash is clearly present in the first edition, with its distinctive, almost round, shape. It is joined in subsequent editions by Victoria Park in Bow (opened 1845), reflecting the Victorians' new interest in health and exercise. Later editions also reveal the creation of London's new centrepiece, Trafalgar Square (completed 1845), with both Nelson's Column and the new National Gallery (1832–1838) constructed on the site of the former royal mews.

The Greenwoods and others benefited from the reluctance of the new Ordnance Survey, founded in 1790, to tackle the more densely populated areas of London. The new organisation was perhaps apprehensive about taking on to take on the private sector competition, and Ordnance Survey maps of London only started appearing in 1866 and covered the city in its entirety by 1872. In the interim, with its high level of accuracy, the Greenwoods' map was a good substitute and shows in great detail the city in the throws of furious and sustained development and change. It was also one of the last major map series to be labouriously coloured in by hand. Soon colour lithography would take over and the natural beauty of the applied colour wash would disappear altogether.

By the turn of the twentieth century, London's inhabitants and visitors had a wide range of maps from which to choose, from those produced by the Ordnance Survey to one of the many popular and cheaply printed colour maps or a printed book of street maps such as *Collins Illustrated Atlas*, first produced in 1854. All of them showed a London apparently bursting at the seams. Apart from the various parks and the familiar profile of the river running across the centre of the city, blocks of development would fill almost every area of the sheet or page.

Map of London from an actual Survey made in the years 1824, 1825 and 1826, 1827
Christopher and John Greenwood

The accuracy of this large-scale map, made by the Greenwood Brothers in 1827, follows in the wake of the first Ordnance Survey maps from 1805–1822 that covered the Greater London area. At approximately eight inches to one mile, the Greenwoods considerably outdid the Ordnance Survey in terms of detail and made any further work on the London area unnecessary for several decades.

The Greenwoods' map coincided with an outbreak of city re-planning, including the new Regent's Street running from Oxford Circus to Piccadilly Circus and the beginnings of clearance of the Queen's Mews at Charing Cross to create Trafalgar Square. Belgravia is located in the middle of development while Marylebone is largely complete on the land of the various aristocratic and school estates.

The stunning hand colouring of the map particularly brings out the river and other bodies of water scattered around London: the reservoirs in Hyde and Green Parks, the lakes in Regent's, Hyde and St James' Parks and the new dock areas in the East End. The essentially frivolous water features of the West End and the hard-edged working docks in the east are the clear marks of a divided city.

Image courtesy of the Guildhall Library, City of London.

Bauerkeller's New Embossed Plan of London, 1841
G Bauerkeller

Right:

Cross' New Plan of London, circa 1850
Joseph Cross

Baurkeller's map celebrates the arrival of colour printing, in no uncertain terms, and uses the newly available technology to achieve a tactile quality in its representation of the city. Different jurisdictions are also boldly colour-coded, giving London the feel of an artificially dyed microscope slide.

The flamboyant graphic and searing colour of this map was not going to encourage others to copy its techniques, however the ability to print in colour would make a great difference to mapmaking from

then on. The choice and positioning of colour would no longer be the responsibility of the colourist. It became part of the standard toolset available to the cartographer, allowing a far wider range of information to be added, which mapmakers did not hold back from providing.

Image courtesy of Cambridge University Library.

This map is a late edition of one originally published in 1828 and regularly revised and reprinted in 13 subsequent editions until circa 1864. In this edition, Victoria Park in Hackney and the new Houses of Parliament were added.

Image courtesy of David Hale/MAPCO.

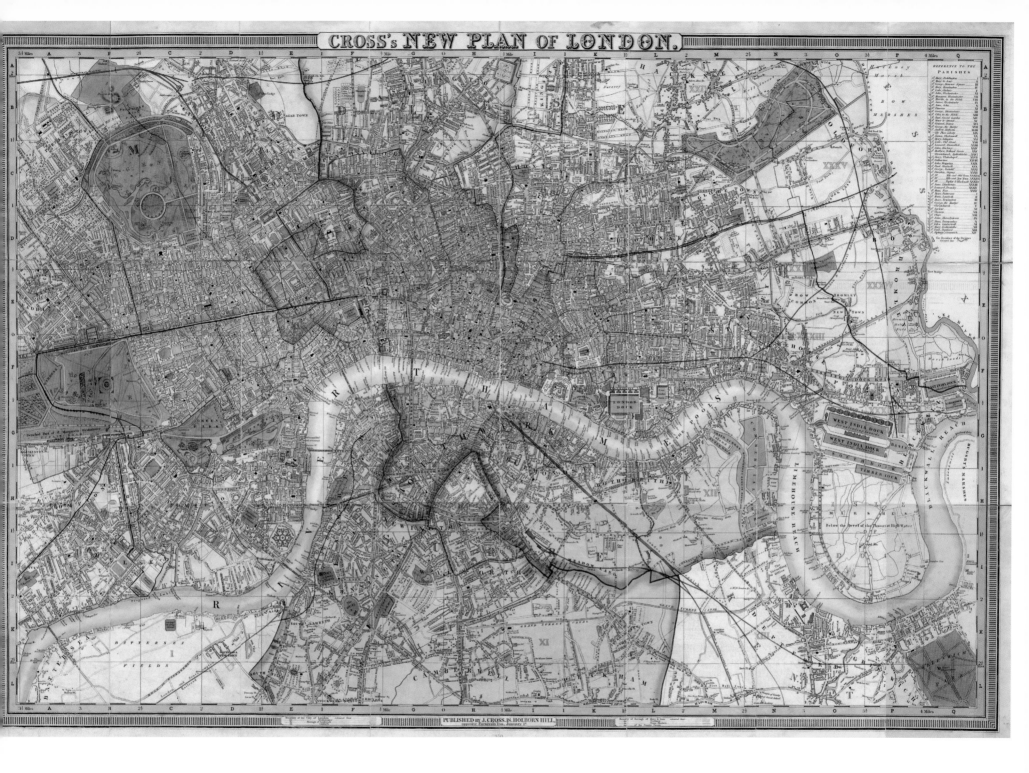

CROSS's NEW PLAN OF LONDON.

Whitbread's Map of London extending four Miles round Charing Cross, 1865
Josiah Whitbread

Whitbread's map readily shows the speed of development engulfing London at this time. On maps of only 15 years earlier, Victoria Park was still set among fields and villages, but here it is enclosed by development on three sides. Similar development is wrapped around Hyde Park and Kensington Gardens while streets of houses are proliferating to both the north and south.

Image courtesy of David Hale/MAPCO.

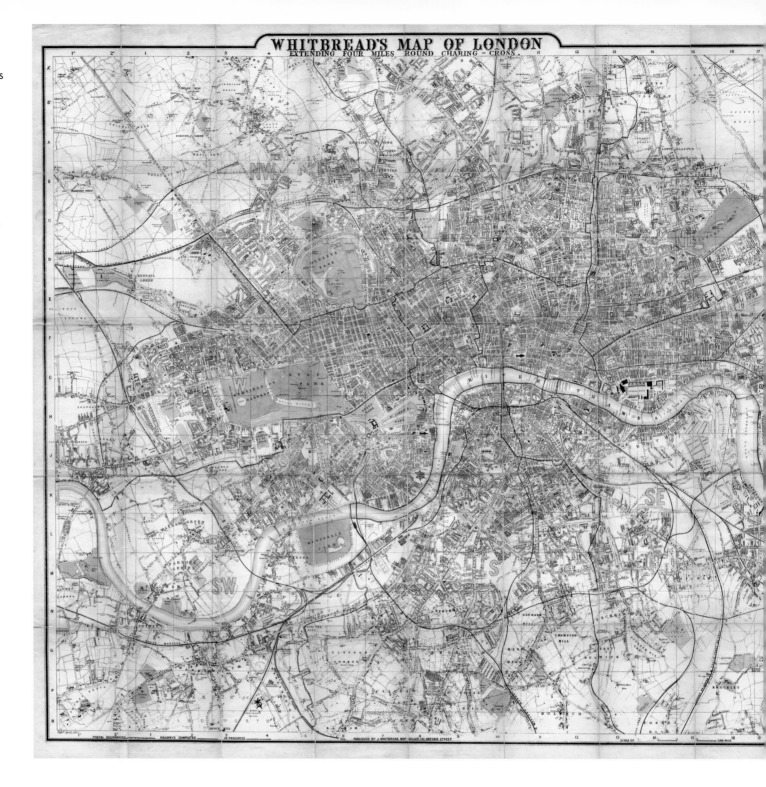

Stanford's Map of Central London, 1897
Edward Stanford

Edward Stanford printed his first authoritative map of London in 1862. At this time he was one of the most prominent mapmakers in London. At the end of the nineteenth century, the population had risen to six million and so the streets are depicted as a dense tangle of development pushing the geographic centre of London to the west—so much so that this map could declare itself central without including anything of the loop of the Thames around the Isle of Dogs.

Like most of Stanford's output this is a serious map, dedicated to accuracy and high levels of detail. It does not seek to use graphic devices to emphasise or exaggerate particular features or types of information. It has an approach that values knowledge for its own sake and, to an extent, marks the end of an era of maps—as the colour printing technique freed mapmakers from the constraints of awkward craft methods. From this point, maps could be drawn

directly onto paper, or film and photography used to transfer the separate colours to the zinc printing plates. It changed the possibilities for mapmakers and brought in a very different style—one more in tune with the aspirations of the new century.

Image courtesy of David Hale/MAPCO.

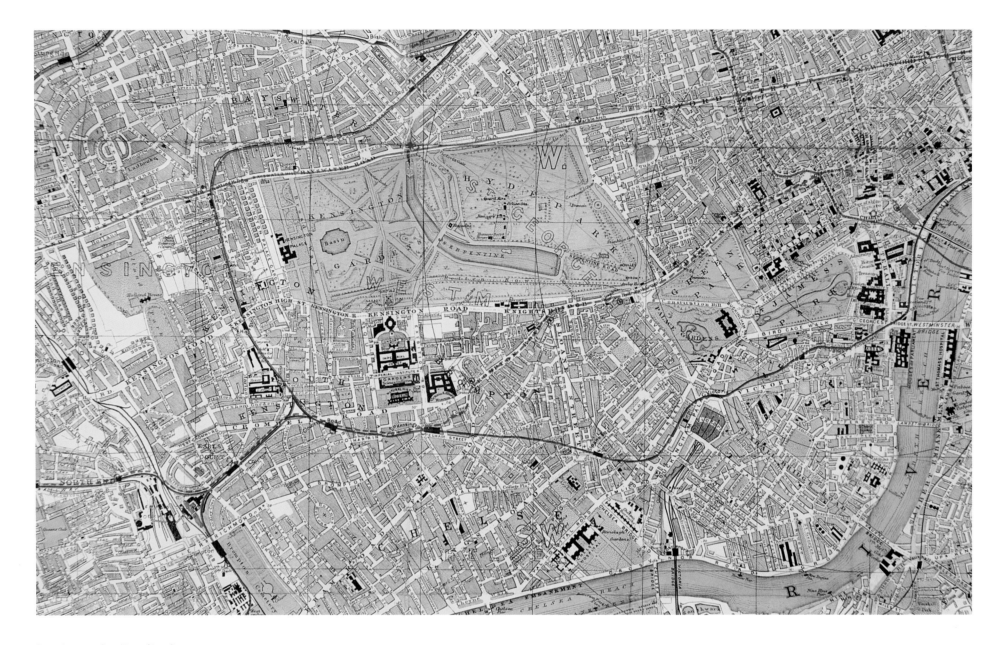

Bacon's up-to-date Map of London, 1907
George Bacon

George Bacon's map concentrates on the fully developed areas of London, depicting a city that has no room for growth. Within the frame, the city is thriving, highlighted by the injection of colour and detail made possible by the new printing techniques. London Underground lines are shown vividly coursing through the city—connecting one part with another rather than bringing in sustenance from outside. It is an optimistic vision of London at a particular moment in the city's history as the Victorian era came to an end and a new generation prepared to take over.

Image courtesy of the Guildhall Library, City of London.

New Library Map of London and its Postal Areas, detail, circa 1930
George Philip and Son

By 1930, London was approaching its peak of population and planners were in urgent discussion about how to limit the growth of a city whose expansion seemed unstoppable. The postal boundaries on Philip's map had been superceeded as had the other significant Greater London zone at this time—the area covered by London Passenger Transport Board. The idea of creating a Green Belt around the city, in order to control its sprawl, was first propounded by Herbert Morrison in 1934 and became law in 1938.

Philip's map, through the superimposition of the postal districts elegantly expresses the growth as if by a series of annular rings, before any of the ring roads that now define London were developed.

Image courtesy of the Guildhall Library, City of London.

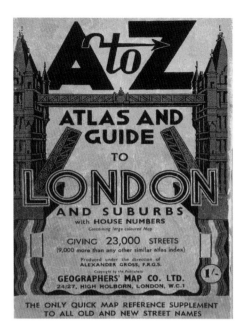

A–Z Atlas and Guide to London and Suburbs with House Numbers, 1936
Phyllis Pearsall, Geographers' Map Company

The origins of the *A–Z*, and the Geographers' Map Company that emerged from it, have been given the gloss of legend in the story of penniless and hungry Phyllis Pearsall walking every street of London to map the capital at the height of its growth and population. The story may be essentially true, but Pearsall was the tough daughter of the map publisher, Alexander Gross, and she used his best cartographer, James Duncan, to draw up the pages for her guidebook using information from the Ordnance Survey. The *A–Z* undoubtedly had a difficult birth and only just survived wartime security restrictions and paper rationing. Having done so, it consequently emerged as the market leader in the late 1940s.

The *A–Z* is in a long line of street atlases produced for London but it achieved the rare distinction of becoming a generic household term and defined the very nature a what a street atlas should be.

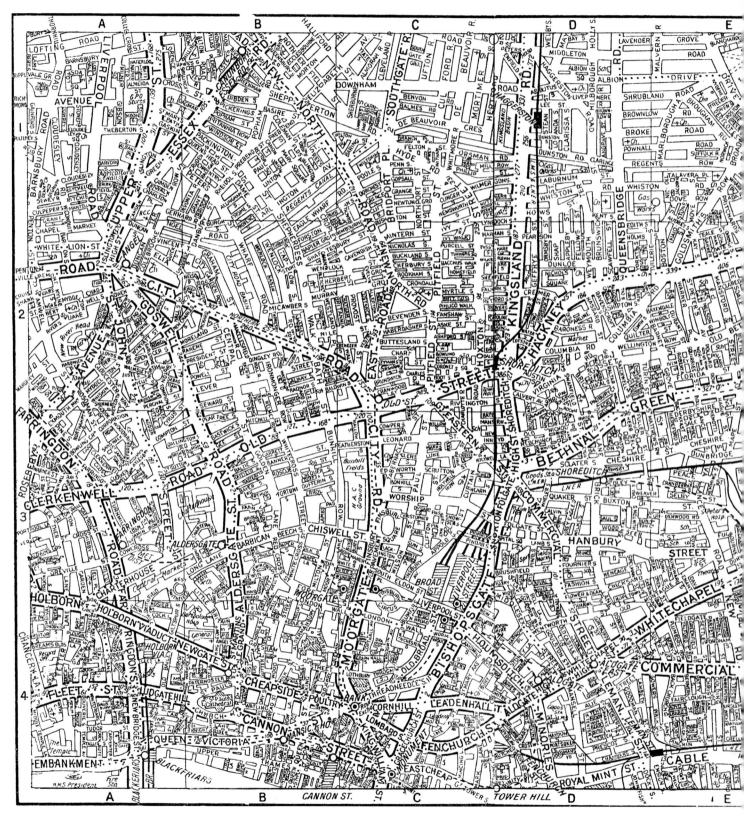

The population of Greater London had exploded from just under one million, during the nineteenth century, to over 6.5 million. On Bacon's map of 1902, covering an area approximately ten miles east to west and eight miles north to south, only the top left hand corner around Willesden and Cricklewood is still largely white and ready for further development.

Dominating all turn-of-the-century maps—and for the first time—is the transport system: trains, Underground lines and roads. Popular maps had become, above anything else, a means of navigating across an increasingly complex metropolitan area and were intended to accompany the traveller on their journey, whether in folded or in book form. To many, London had succumbed, in Henry James' description, to "this senseless bigness".[3] It was doubtless also a frightening place where fear of getting lost, or becoming prey to the bestial criminal class luridly publicised in papers such as the *Pall Mall Gazette*, was rife. A map could be a talisman in such a dark and dangerous place—an object of order in a place of chaos.

What maps of this era don't readily show is other infrastructure efforts, Joseph Bazalgette's and the Metropolitan Board of Works' (MBW) embankments and drainage works, the water supply networks, and the upheaval caused by positioning railway lines so close to the centre of the

city, for example. But the evidence is there, in the narrowing of the Thames and creation of the embankments, and the disappearance of the worst of the slums such as the Nichol in Shoreditch or Agar Town on the route of the lines into St Pancras Station. Similarly, New Oxford Street drove through the St Giles rookery ostensibly—an attempt at useful road planning but also dealing a convenient blow to the most notorious slum in London.

At the start of the twentieth century, London had a new self-image to project. As a city, it was now the capital of the largest empire the world had ever known and yet, despite all the building work carried out during Victoria's reign, it did not appear willing to play the part of the great Imperial city. It had no set piece of grand urbanism to compare with other aggrandising capitals and was easily outshone in that respect by Berlin, Paris and Washington. The result was two great bombastic schemes; the connection of The Mall, which had previously terminated in a dead end, via Admiralty Arch, designed by Aston Webb, 1908–1909, into Trafalgar Square, and the carving out of Kingsway and the crescent of the Aldwych from Holborn down to the Strand. They were both self-conscious attempts to make a mark on the map of London, but neither caught the spirit of London, and future re-planning exercises always came with a more practical purpose if not always less destructive results.

In 1935, Phyllis Pearsall began a personal survey of London. Walking over 3,000 miles, down 23,000 streets, culminated in the creation of the *A–Z Atlas and Guide*. A great deal of myth making has surrounded this relentless journey, including rumours that no street atlas had existed before the *A–Z*. In truth, books of street maps had existed for decades, including the Collins map of 1854 and even one produced by Pearsall's own father, Alexander Gross of the Geographia Company, in 1913.

The *A–Z* was undoubtedly more useful and up-to-date than previous attempts, but it was hardly revolutionary, and its main practical advance was the inclusion of house and building numbers. By the 1760s, house numbering had become standard practice in London and numbers were regularly marked on maps including Richard Horwood's of 1792—so not even this was new. The real difference the *A–Z* made was in its simplicity and ready availability. It was far from a precious item, cheaply printed in black and white and happily abusable. Perhaps because of this, and once Pearsall's Geographers' Map Company had survived wartime restrictions and paper shortages, it became a household item available to almost anyone.

The ubiquity of the *A–Z* made navigating strange and new areas of London a practical and straightforward possibility. If public transport, and especially the London Underground system, made it possible to reach the general vicinity, the *A–Z* took you to the right door on the right street. It made all of London accessible. If you were lost, it could tell you where you were and how to find your way again. Despite its initial crude and make-do cartography, it perfectly served its purpose, as it continues to do—making it a strong candidate as one of the key maps of London.

London in the 1930s reached the height of its population, even after the death toll of the First World War. The numbers in the inner-city area covered by the London County Council (LCC) decreased slightly, to about four million, but those in the wider area of the Metropolitan Police District exceeded 8.5 million. As these figures indicate, the suburbs had grown substantially. London Underground lines now extended overland, deep into the (now accessible) countryside around the city and the road network, including new arterial roads and the North Circular Road, which had been greatly extended to cope with the transport pressure that a city of such enormous size generated. Motor vehicles, including buses, had taken over the roads and had made the growth of now fairly distant suburbs possible. The relatively compact plan of London's development in 1900 extended out in uncontrolled sprawl down every available transport corridor to a distance of ten miles and more from the centre. Even before the Second World War was over, measures would be taken to control that spread and to reduce London's population.

In the post-war years it had dropped to a low of approximately 6.8 million in 1981. By the millennium, London's population was growing steadily and had risen to over seven million, aided by immigration from both the UK and the rest of the world. The present shape of London has become defined by the Green Belt imposed in the 1938 and also by the M25 motorway circumscribing her immense growth. But, just as in previous centuries, development has long since leapt over any political or geographical boundaries, with much of the southeast becoming an extension of the city and with over a third of London's workforce daily commuting more than 15 miles to work in the central areas of the city.

Descriptive maps continued to be made of London during the twentieth century but without making any great cartographic strides forward. Advances were made in maps that described other aspects of the city; including social, political, demographic and environmental information and comparisons. These are discussed in subsequent sections of this book.

It is only at the end of the century that another shift took place and with new developments in two separate directions. Hand drawn maps had become a rarity by the 1980s and 90s and the new computer mapping files were rapidly translated to provide digital and online information on London—initially in a conventional format and then to provide increasingly rich information. Driven in part by the needs of technology providers for content, but also by a perception that interactive mapping could become a good sales and marketing tool, it has initially been popular for that most conventional use of maps—navigation.

Route finding systems plot journeys from location to location and provide maps and advice on potential modes and routes. Combining mapped information with global positioning systems make it possible for devices to respond to actual movement within the virtual landscapes on view. Real-time directionality makes the old conventions of placing north upwards unnecessary as the map swings around with

West End Map of London, 2006

Silvermaze Ltd

From the twenty-first century, map production has been revolutionised by the computer (as much as they were at the start of the twentieth century by photolithography). One of the results has been the return to bird's-eye isographic maps showing each individual building block—often in a cheerful, cartoon-like style. One of the leaders in this field has been Silvermaze who use their adaptable virtual version of London to generate bespoke maps for a range of uses and customers.

the reader's viewpoint, imitating generations of confused map users. Such satellite navigation devices have little interest in the conventional flat ground plan map-view. A perspective view of the streets ahead only, has become the preference and is likely in turn to be replaced by high quality three-dimensional views of environments as satellite generated survey information becomes more available.

In parallel with this new and popular map view, has come another move into three-dimensional representation (the first significant change in over three centuries), this time borrowed from computer games, animation and computer aided design. The graphic style of games like Simcity, for example have developed into mapping tools of London. Companies such as Silvermaze are drawing detailed axonometric maps of London and Local Authorities, including Islington, are commissioning representational three-dimensional 'walking maps'. Perhaps, because computers allow the potential to map in three (or more) dimensions, there has been a natural preference for a more visually friendly style. Such an inclination has resulted in representations of the city not too dissimilar to earlier renderings; the Copperplate, the Agas and Richard Newcourt's of 1658. Has

London mapping simply come full circle or is new topographical and imaging technology about to take us somewhere far more interesting? Photorealistic three-dimensional maps of cities, including New York, being developed by companies such as Microsoft certainly suggest we are on the edge of entering a new phase.

The shape of the map of London in the early twenty-first century is, as ever, one of constant debate. It is anticipated that the greater metropolis will go on expanding for another 2,000 years of almost uninterrupted growth. But there are now many cities much larger than London across the world. Forces such as global warming and restrictions on the use of fossil fuels may require a reassessment of large urban conglomerates and redefine the map of London once again. Whatever it becomes, mapmakers will be there to describe it in new and creative ways and to produce new cartographical markers in its continuing history. The first great map of twenty-first century London is eagerly awaited.

LONG VIEWS AND PANORAMAS

Although the focus of this book is on London's maps, there are certainly other ways to accurately represent a city. The map uses an orthographic projection to set out its image, but trajectories from a single point—as if from a travelling camera—have played a similarly descriptive and informative role.

Long views are a staple of early depictions of London, invariably encouraged by the experience of travelling the Thames and watching the townscape unroll like a Chinese scroll. The earliest example, a sketch by Wyngaerde, circa 1540, adopts a low level view that depicts little more than a few riverfront buildings and the church steeples that poke out from behind them. Later, artists took a more elevated perspective, as if floating above the streets alongside the trumpet blowing angels as featured in Visscher's and Hollar's works. Hollar cites the tower of St Mary Overy (now Southwark Cathedral) as the location of his long views, however, in reality, the vantage point had to be much higher in order to capture the individual building elevations across the river. Artists found no difficulty in imagining a bird's-eye view and faithfully rendering it, but today's historians feign shock that such views have been depicted from locations that the artist had no conceivable access to.

The concept of the panorama, the 360 degree view from a single perspective, came significantly later and was the vision of Robert Barker, who both coined the word and patented the invention in 1787. He was responsible for creating the first such example, a 180 degree view of Edinburgh, which he exhibited in 1788 in the rather cramped conditions of his house in London (which perhaps explains why it was not a great success). Four years later, he returned with the first 360 degree view, this time of London seen from the top of Albion Mills, at the southern end of Blackfriars Bridge. The success of this panorama led to a global craze (christened panoramania) and large rotundas to house this new art form began springing up all over the Western world. Barker toured his London panorama across Europe with stops in Hamburg, Leipzig, Vienna, Paris and Amsterdam. Panoramas were a pre-cinematic sensation and their producers would compete to present the most exotic and crowd-pulling scenes, which included views of famous cities from the grand tour and battlefields, for those who would never see or experience them in any other way. Views of London rarely provided such visceral thrills, but the fame of the city meant that panoramas continued to be planned and produced.

The Tower.

Tower Wharfe

S. Olafe

London Long View, detail, 1647
Wenceslaus Hollar

Wenceslaus Hollar—one of London's greatest
topographical artists—produced innumerable views
across the Thames, including several showing the
Great Fire of London and its aftermath in 1665. The
Long View, drawn and engraved 18 years earlier, is his
masterpiece detailing London from above Southwark.
The buildings and streets of London are detailed
lovingly below and angels play high over the river.

Hollar was a mapmaker as well as the creator
of highly detailed views of buildings (often from
unaccessible aerial views) his records of London
are as much from the perspective of a surveyor as
artistic interpreter so this is a fairly accurate image
of London at the beginning of the English Civil War.
None of the conflict that took place is evident in
Hollar's map. Rather, he presents an image of a
prosperous and peaceful city (quite possibly because
he and his family had left London in 1644 for Antwerp
to escape the war and would not return again until
1652, when stability had resumed).

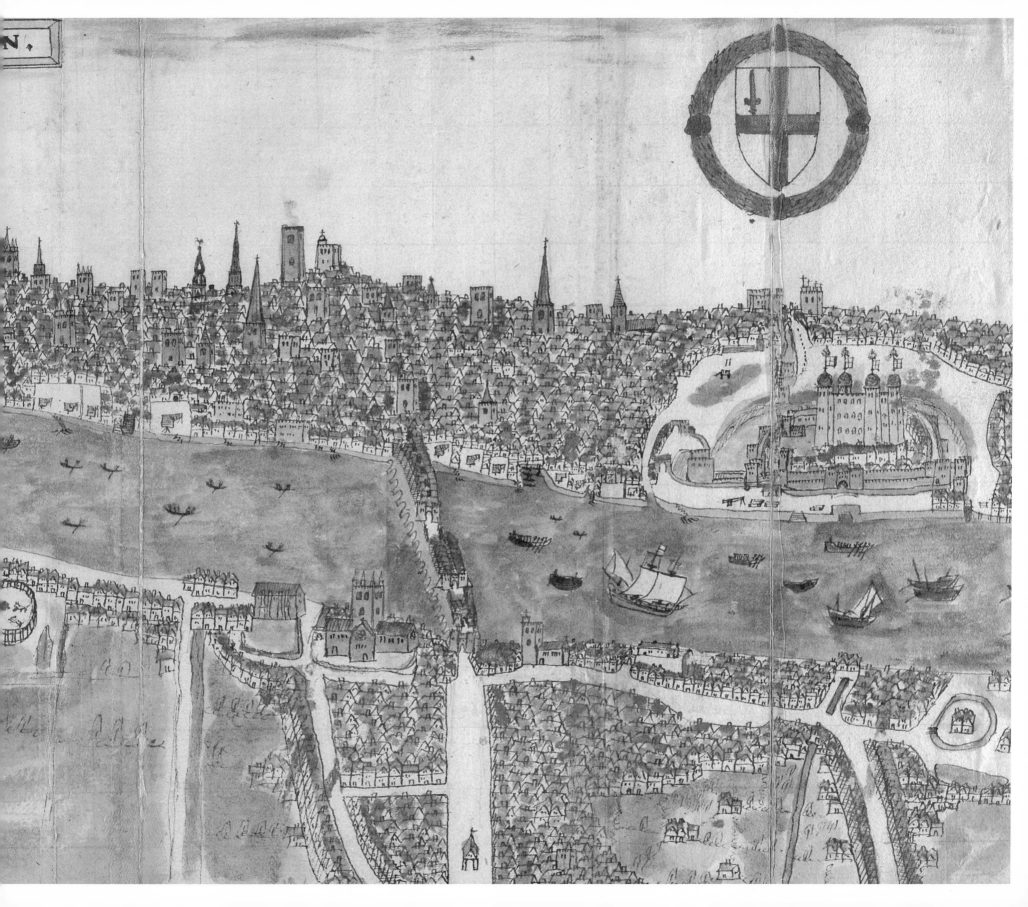

Previous pages:

View of London from the south, showing the River Thames, London Bridge and the Tower
—The Particular Description of England, 1588
William Smith

This busy view of London in 1588 probably derived from the work of other mapmakers (Braun and Hogenberg, for example) and Wyngaerde's Whitehall Palace mural, and demonstrates all the hallmarks of a modern souvenir map.

 William Smith's drawing may not provide more than the most basic of spatial understandings of London but, as with many other topographical images of London, it does provide a real feel for the character of the city, particularly the number of churches that dominated the skyline and the frivolous activity that took place across the river on Bankside (away from the controlling restrictions of the city authorities).

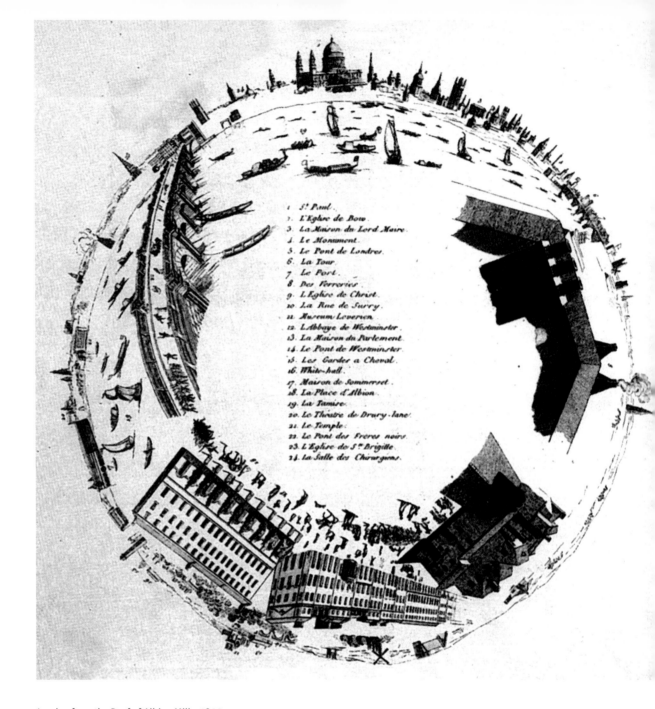

1. St Paul.
2. L'Eglise de Bow.
3. La Maison du Lord Maire.
4. Le Monument.
5. Le Pont de Londres.
6. La Tour.
7. Le Port.
8. Des Verreries.
9. L'Eglise de Christ.
10. La Rue de Surry.
11. Museum Leverien.
12. L'Abbaye de Westminster.
13. La Maison du Parlement.
14. Le Pont de Westminster.
15. Les Gardes a Cheval.
16. White-hall.
17. Maison de Sommerset.
18. La Place d'Albion.
19. La Tamise.
20. Le Théatre de Drury-lane.
21. Le Temple.
22. Le Pont des Freres noirs.
23. L'Eglise de St Brigitte.
24. La Salle des Chirurgiens.

London from the Roof of Albion Mills, 1802
Robert Barker

At the end of the eighteenth century, Robert Barker, Edinburgh painter, entrepreneur and showman, developed the idea of the panorama. He toured this panorama of London—a 360 degree view of the city from the roof of a building at the southern end of London Bridge—across Europe and this image depicts the explanatory card given to audiences at this time. Barker went on to produce scenes of more exotic places and events to satisfy his audiences—for many this would often be the only opportunity to experience a foreign city.

These included two watercolour paintings made in 1820 for conversion into a panorama in America (the Rhinebeck panoramas) and two views from the cupola of St Paul's, the first painted by Horner in 1829, with a second following in 1845.

Later in the nineteenth century, the invention of the hot air balloon led to new opportunities for descriptive artists to show off their abilities. Unlike the imaginary elevated views of the seventeenth and eighteenth centuries, it was now critically important to have been there and had the—quite possibly terrifying—aeronautical experience. The images that resulted from these expeditions showed the curvature of the earth and copious clouds conveniently obscuring areas of the city. The master of this form of reportage was JH Banks. Others who took the opportunity to view London from such heights included the writer and social reformer, Henry Mayhew, who wrote in 1862: "To take as it were an angel's view of that huge town where, perhaps, there is more virtue and more iniquity, more wealth and more want, brought together into one dense focus than in any other part of the earth."[4]

The balloon view excited and inspired a new perspective of London. However, by the turn of the century, its heyday had passed.

Such a format was only briefly returned to in the twentieth century, when Cecil Brown created his 1945 view of a war-damaged London as surveyed from a barrage balloon.

The panorama looked even less likely to have a revival. However, computer-imaging techniques have recently revisited the concept and 360-degree photographic panoramas are now available from an eclectic variety of locations including all London's bridges and many other sites. They have also been produced to illustrate the impact of proposed new buildings on the skyline of the city, resulting in new high-level panoramas that show the anticipated perspective from such spaces. Such elevations can be found, for example, in the projected view from the 310 metre high Shard Tower. To be built at London Bridge, the Shard will be located immediately next to the position of Albion Mills where Robert Barker had conceived his London panoramas in 1792. It would have been good to see it in Barker's Leicester Square Rotunda—the nineteenth century crowds would have loved it.

Rhinebeck Panorama,
circa 1810

Discovered lining a barrel of pistols in Rhinebeck, USA, in 1941, the four large watercolour drawings were apparently intended for conversion into full-scale panorama paintings in America. The views show a Thames crowded with shipping and an elegant, almost classical city, beyond. The panorama clearly presents London, but it is an interpretation that looks more like Stockholm or St Petersburg than the chaotic nineteenth century Medieval city of reality.

**A view of the St Paul's Panorama,
Regent's Park, 1829
Thomas Horner**

From 1824–1829 Thomas Horner had a purpose-built rotunda, The Colosseum (but more closely resembling the Pantheon in Rome) constructed in one corner of Regent's Park to a design by Decimus Burton. The building was to house his panorama of London that he had drawn from a scaffolding crow's nest on top of the cupola of St Paul's Cathedral in 1821 and then painted by ET Parris on a series of vast canvases ("more than an acre") that were stretched around the walls. Viewers entered an 'ascending chamber' that was mechanically raised in a central drum to two viewing galleries that mimicked the galleries above the Cathedral dome.

The image is a painting of The Colosseum's interior, showing both the central viewing tower and the panoramic painting beyond. As a result of the popularity of Horner's panorama, it was repainted 16 years later. London was finally replaced by a view of Paris in 1848 and The Colosseum was put up for auction in 1855 and demolished 20 years later.

A view of London and the surrounding Country
taken from the top of St Paul's Cathedral,
circa 1845

A souvenir, perhaps, from the Regent's Park Colosseum,
to celebrate the repainting of the London panorama.
This print was intended to be viewed with a mirror
polished cylinder placed in the centre.

Following pages:
A Panoramic View of London, 1845
JH Banks

This panoramic view both celebrates the
connectivity of the area around St George's
Circus and Elephant and Castle (as a result of the
proliferation of new bridges across the Thames),
as well as the dark and glowering nature of London
itself. Banks took to a hot air balloon to achieve
this view of London subsumed in the shadows.
Such an impression provides a more accurate feel
for London at this time, a city both of industry
and great poverty. Banks undoubtedly expresses
the general feeling at this time that the growth of
London was never ending. The six bridges shown
are: Westminster (1750, shortly to be replaced
in 1862), Hungerford footbridge (designed by
Isambard Kingdom Brunel, 1845, replaced by the
Charing Cross Railway Bridge in 1864), Waterloo
(conceived by Sir John Rennie, 1812, and replaced
in 1942), Blackfriars (planned by Robert Mylne,
1769, and replaced in 1869) and London (designed
by Sir John Rennie, 1831, replaced in 1972 and
now re-erected in Lake Havasu, Arizona).

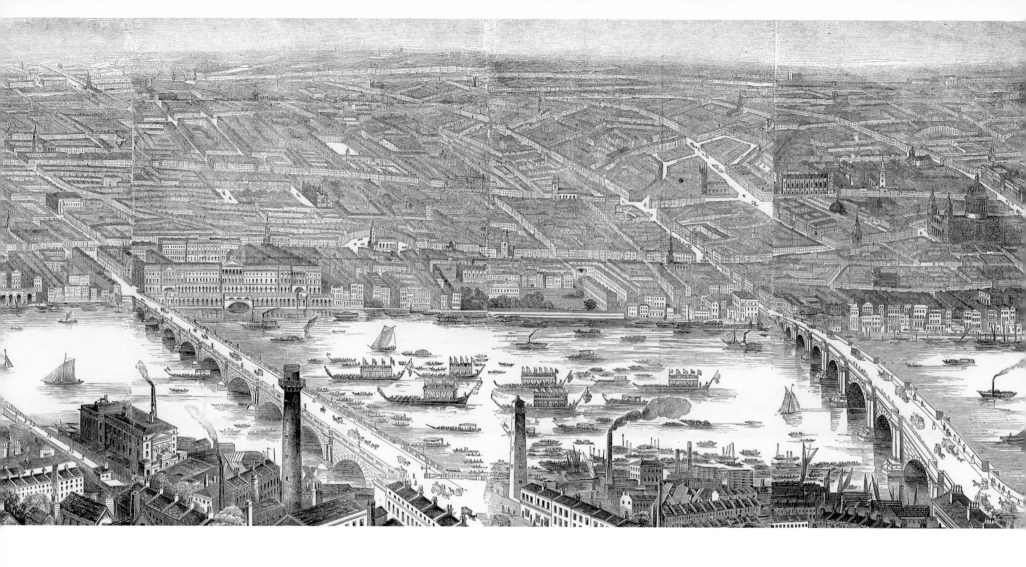

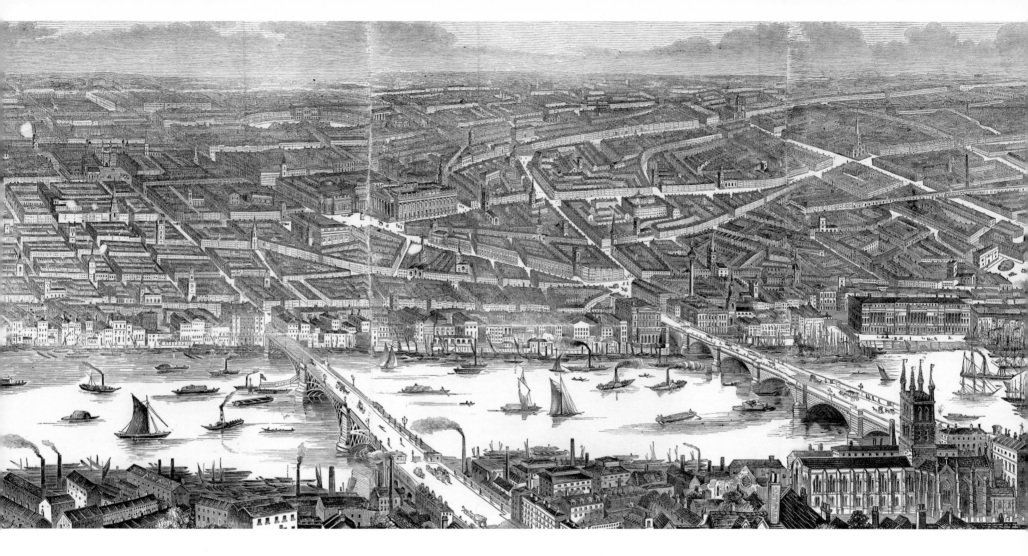

Panorama of the River Thames in 1845,
supplement to *The Illustrated London News*,
11 January 1845
SC Smyth

This image resulted from a publishing war between the *The Illustrated London News* and the *Pictorial Times*—then, as now, fought with free gifts. In 1842, the launch issue of the *The Illustrated London News* was accompanied by the offer of a copy of "the splendid Colosseum Print of London" to everyone who took out a six month subscription. This first print appeared in 1843 but, by 1845, its engraver had become the publisher of the new *Pictorial Times* and was offering "for all persons, the largest engraving in the world, the Grand Panorama of London from the Thames, fourteen feet in length". *The Illustrated London News* retaliated with the panorama illustrated, issued on the same day in 1845. It shows an elevated view of London depicting the city across the Thames from the south.

Image courtesy of David Hale/MAPCO.

**Projected Panorama view from the
Shard Tower, London Bridge, 2005
Hayes Davidson and John Maclean**

The reinvention of the panorama resulting from
digital photography and image-making is one of the
unexpected oddities of the digital age. This image,
a perspective from the top of the yet to be built
Shard Tower over London Bridge Station is just as
interpretative as Hollar's view.

© Sellar Property Group.

Demographics

Throughout the nineteenth century, maps began to develop in new ways. Instead of simply describing what was at ground level, or being useful tools to help navigate from A to B, they began to reveal other phenomena and to uncover and explain patterns in the city that were invisible to the naked eye.

Possibly the most famous of these maps is one by Dr John Snow, which depicts the distribution of cholera around the Broad Street pump in Soho. This map, part of the evidence he compiled to prove his hypothesis on the water-born nature of the disease, appears to be straightforward, although relatively few people were convinced of this at the time. Snow's map now represents the acknowledged beginning of the modern discipline of epidemiology, but is also a talisman for many other scientific and social studies that use mapping techniques to reveal truths that are almost impossible to identify otherwise. Like aerial photography—when used in archaeology to reveal long buried structures and traces of activities underlying landscapes—plotting information on maps can make immediate graphic sense of apparently unrelated information. As a result, it became essential for innovative mapmakers, and the new breed of social scientists, recorders and campaigners, to find new ways of presenting freshly collected data.

Well before Snow, other disciplines were revealing how such data could be presented. In 1800, Thomas Milne adapted his survey information of the Cities of London and Westminster to show the variety and pattern of land use, using colour and hatching to reveal different agricultural uses and building ages across a wide area around the city. At the same time, William Smith was sketching out the first incarnation of his geological map of England (although he only produced the fully printed and coloured version in 1815) that showed the underlying structure of rocks and strata beneath the soil. Depictions of the London and Thames Valley area itself did not emerge until 1851, when Joseph Prestwick produced the first hydro-geological map in *A Geological Inquiry respecting the Water-Bearing Strata of the Country around London*, and later in 1856, when Robert Mylne turned his contour mapping work for water companies into a graphic study of the rocks below.

It was in the mapping of population that the new techniques of Geographical Information Systems (GIS) came into their own.

**Population density of London and
the surrounding Counties extending
for about 50 Miles round, 1891
John Bartholomew**

One of the earliest maps showing population densities
plotted with isarithms or 'contours' of equal population
density, Bartholomew's map gives an immediate visual
representation of the inhabitation of an area. Statistics
on population had been gathered for many centuries,
but the use of maps to represent this data was a new
departure and would prove immensely valuable in the
years to come.

While there are no major surprises concerning the
populous of the city centre at the heart of this map,
the spikes that follow significant valleys, roads and
railways out of London are illuminating. In many
ways, the true shape of London is the area in grey—a
highly distorted shape that pulls the city off to the
west and southwest and also northward along the
Upper Lea Valley.

Image courtesy of the Guildhall Library, City of London.

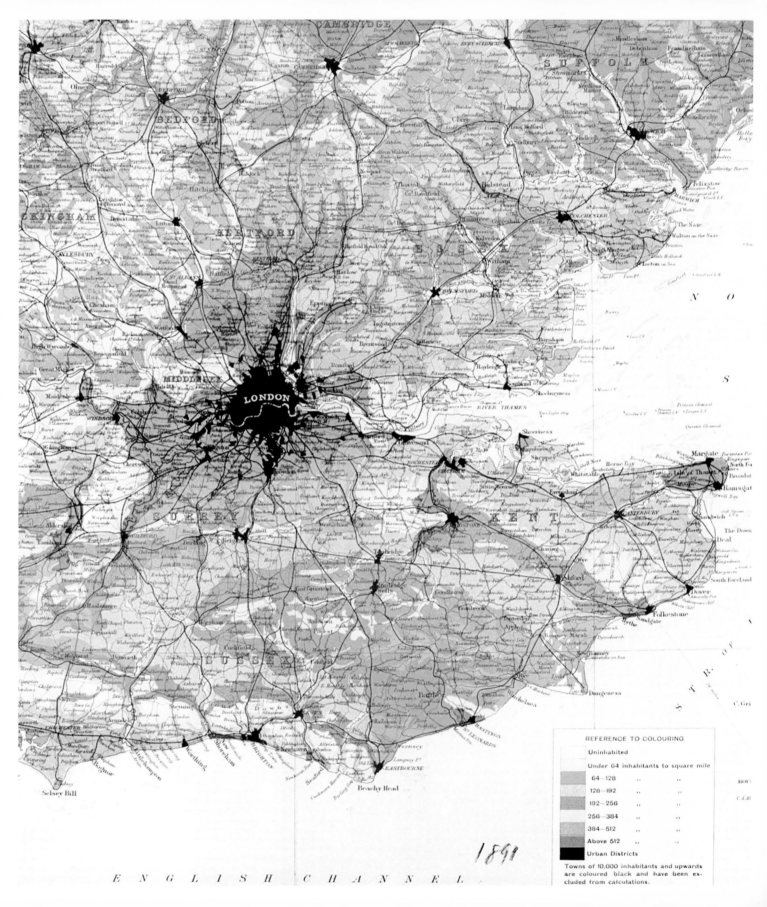

REFERENCE TO COLOURING

Uninhabited
Under 64 inhabitants to square mile
64—128
128—192
192—256
256—384
384—512
Above 512
Urban Districts

Towns of 10,000 inhabitants and upwards
are coloured black and have been excluded from calculations.

JEWISH EAST LONDON

SCALE

This Map shows by Colour the proportion
of the Jewish population to other residents of East
London, street by street, in 1899.

EXPLANATION OF COLOURING.

Proportion of Jews indicated.

95% to 100%.

75% and less than 95%.

50% and less than 75%.

25% and less than 50%.

5% and less than 25%.

Less than 5% of Jews.

NOTE.—In all streets coloured blue the Jews form a majority of the
inhabitants; in those coloured red, the Gentiles predominate.

Jewish East London in 1899, 1901
George Arkell

A byproduct of Charles Booth's survey activity,
this map of the Jewish population in east London
highlighted the great number of Jewish people living
in the Whitechapel, Spitalfields and Mile End areas
of Stepney. Arkell prepared the map for publication
in the book *The Jew in London* (by C Russell and HS
Lewis) under the auspices of the East End centre for
social reform, Toynbee Hall. The number of Jewish
people in the area made up only 18 per cent of its
inhabitants, however, the map's colour-coded nature
gave a different impression and it became a weapon in
the hands of xenophobes. As with other plans emerging
from Booth's work, maps showing population shifts
and origins are still actively produced today—just as
statistics on immigration continue to be abused.

Image courtesy of the Guildhall Library, City of London.

People and populations had been invisible on maps of the city since the very early maps of the sixteenth century. Now, their presence would be felt through the use of colour-coding and elaborate figures. In 1891, in a similar vein to the geological maps of earlier in the century, Bartholomew produced a map of the southeast of England, in which he presented densities as contours that were colour-coded in between. The interpretation of statistical information on maps—in the guise of landscape relief—would prove to be one of the most fruitful and enduring of metaphors for social cartographers and enable them to visualise numerical data as three-dimensional shapes and patterns.

The maps produced by the other great population survey of the time, *The Descriptive Map of London Poverty* by the Liverpool industrialist Charles Booth, began with the publication of the first volume, *Labour and Life of the People*, 1889, which focused predominantly on the East End. These maps were less concerned with abstracting information into broad patterns than recording precise levels of information for each street, by using a version of Stanford's Library Map as a base. Nonetheless, the mass of detailed colour-coded information easily reveals the larger picture and shows the general and characteristic mix of class and wealth in London as well as the poverty of large areas in the East End. Booth and his researchers initially investigated the poverty levels of individual families from School Board records. They assigned each street one of seven social condition classes. These were then coloured from black ("lowest class, vicious, semi-criminal") to yellow ("upper middle and upper classes, wealthy") on the maps.[5] Subsequent surveys updated and extended the range of information gathered and recorded, and Booth published his findings in three separate editions, including more maps, before 1903. The detail of the mapping, and accompanying written material, must have been a key resource for historians subsequently and may, in part, explain the mass of books available that focus on the capital at the turn of the nineteenth century.

The mapping of the demography of London has kept mapmakers busy ever since. Regular census data has assisted this (eliminating the need to follow policemen on patrol in order to gather information, as one of Booth's researchers, George Duckworth, did in the years between 1897 and 1900). By 1974, the authors of *A Social Atlas of London* had clearly spent so long disassociated with 'the streets' they felt it necessary to thank the photographer "who braved two raw November days to take photographs for the cover".[6]

Demographic cartographers throughout the twentieth century restlessly experimented with new ways of presenting ever-growing amounts of data. The compendious *Atlas of London*, published by Emrys Jones of the London School of Economics in 1969—using information from the 1961 census and the second *Land Utilisation Survey*, 1963–1968—found numerous new ways to visualise the data and make it immediately clear and graspable.

The *Atlas of London* tackles a wide range of statistical data, including tenure, land-use, car ownership, immigration and gender and age splits. In its willingness to deal with such large data sets it led the way into the age of computer generated maps. *A Social Atlas of London* was able to repeat many of its charts using specially written computer programs, but this time including 1971 data. The producers of maps showing different aspects of society and demography have never looked back, and relatively crude maps—all with the familiar outline of the London area—can now describe almost any characteristic that has been recorded. What is missing from many of these renderings is the

multi-dimensional quality that was required of their predecessors, and they are generally impoverished as a result.

However, computer technology has brought other advantages, including the ability to render statistical landscapes in three dimensions and to capture information on elusive issues such as rates of change alongside current levels of population. Computer graphics are still finding their way in cartography, and fascinating, often beautiful, experiments are taking place when communicating complex statistics succinctly. Some of those experiments are shown here alongside the more prosaic public information diagrams.

The Descriptive Map of London Poverty, 1889
Charles Booth

The decision by Charles Booth to use maps to illustrate his *Inquiry into Life and Labour in London*, 1886-1903, was momentous for the future of cartography as it was for social science. His inquiry, though detailed and meticulous, was not so unusual and followed work done by Friedrich Engels, Henry Mayhew and many others before him. However, the maps provided the means to communicate his findings immediately and with maximum impact. The data collected in the first study into poverty was straightforward enough, even if his classifications of households into categories were more incendiary.

As his maps of London were published, areas of the city became not only categorised by reputation and anecdote, but by hard facts and science. In some areas, his findings could not have been more damning and must have informed subsequent slum clearance programmes.

Later studies in the programme included research and analysis into living and working conditions, the lives and employment of women, the organisation of trade and industry, the effects of national and international migration and the religious life of the capital. Booth's maps were the forerunners of a new industry mapping social patterns and projections that are now proliferating more than ever.

Image courtesy of the Museum of London.

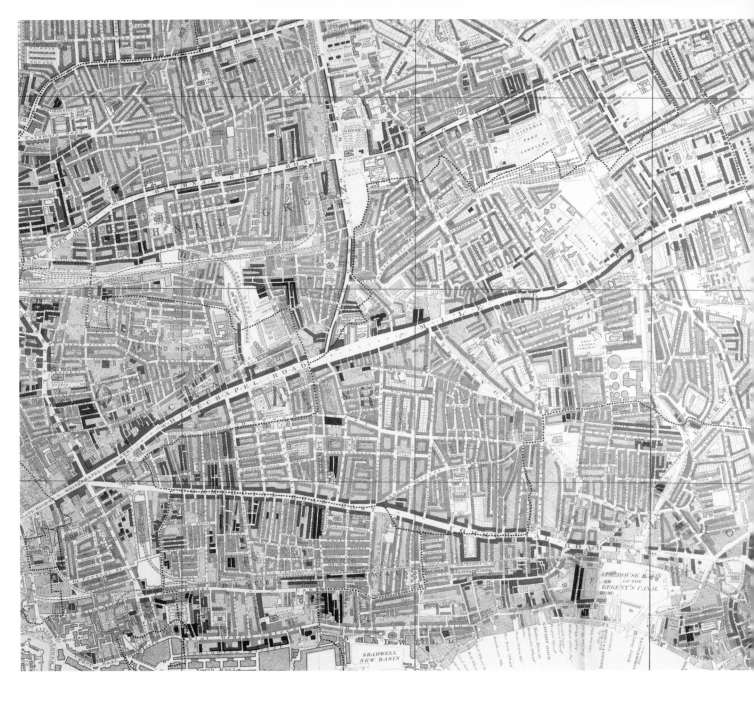

Top:

Diagram of tenure of households from *A Social Atlas of London*, **1974**

Bottom:

Diagram of age/sex structure by London Boroughs from *A Social Atlas of London*, **1974**

John Shepherd, John Westaway and Trevor Lee

A Social Atlas of London published a wide range of maps, mainly based on 1971 census information and following on from pioneering work done by the London School of Economics (LSE) under Emrys Jones in 1969 (*The Atlas of London*). The difference between the two publications is the far greater ease with which the researchers for *A Social Atlas of London* could translate the data using computers into a mapped result. Such transfer of data into graphic form has now become almost automated and is undertaken by research groups at a prolific rate.

Images courtesy of Oxford University Press.

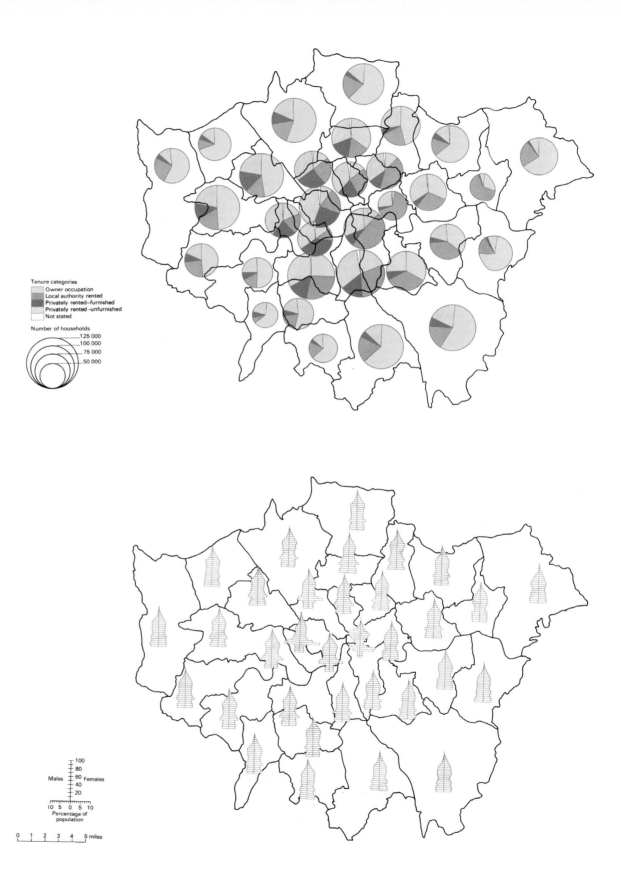

Tenure categories
- Owner occupation
- Local authority rented
- Privately rented—furnished
- Privately rented—unfurnished
- Not stated

Number of households
- 125 000
- 100 000
- 75 000
- 50 000

Males   Females
100
80
60
40
20

10 5 0 5 10
Percentage of population

0 1 2 3 4 5 miles

Atmospheric Disturbances
from *The Weather Book*, 1863
Robert Fitzroy

Fitzroy was a naval captain and, later, admiral.
Accompanied by Charles Darwin, he captained HMS
Beagle on the trip to Tierra del Fuego. It is his post as
the first Meteorological Statist to the Board of Trade
(heading a department that was later to become the
Meteorological Office) however, for which he is best
remembered. On board the HMS Beagle he developed a
'storm glass', a form of barometer that was later widely
distributed. His professional interest in meteorology and
his naval background encouraged Fitzroy to develop
means of "forecasting the weather"—as described in *The
Weather Book*—that were well in advance of the scientific
opinion of his time. This map is missing the isobars
familiar on today's weather charts, and shows a storm
brewing over London and most of the southeast.

Image courtesy of Science & Society Picture Library.

CLIMATE AND ENVIRONMENT

In December 1802, a young pharmacist/chemist called Luke Howard presented a paper to the newly formed Askesian Society, a small group of dissenters and friends fascinated by the new 'rational entertainment' of science. The Society, which met in Howard's employer's laboratory just off Lombard Street in the City, worked on the principle that all its members should read a paper or else pay a fine. When Howard's turn came, instead of discussing chemistry, he chose to address the subject of clouds—a private enthusiasm rather than a professional interest—inspired by the view from his window in Plaistow (and perhaps by the recent activities of various intrepid and foolhardy balloonists). His paper, "On the Modifications of Clouds", which classified cloud formations along Linnaean lines, was published almost immediately and became the founding text of the new science of meteorology. It also led Howard, later to be elected to the Royal Society, to start studying and mapping *The Climate of London*, the subject and title of the substantial book of collected writings he published in several volumes from 1818 onward.

In this volume, Howard published charts—if not strictly weather maps—showing London's climate in a format easily recognised today. In doing so, he started the graphic description of London's environment that now forms a substantial part of London mapping activity. He was also the first to describe what is currently known as the "urban heat island effect" when he wrote in *The Climate of London*: "But the temperature of the city is not to be considered as that of the climate. It partakes too much of an artificial warmth, induced by its structure, by a crowded population and the consumption of great quantities of fuel."[7] This characteristic effect identified by Howard was only given the label "heat island" many years later when a map showing isotherms over an urban area was perceived to resemble the contours of a small island rising from the sea.

London has had its own distinctive environmental conditions for many centuries, resulting from both the sheer quantity of fuel burnt, as noted by Howard, and the high sulphur content of the poor quality coal used. In 1661, John Evelyn published *Fumifumigium* in which he complained of the "hellish and dismal cloud of sea-coale".[8] The heavy, thick, smoke-laden fogs that occasionally became trapped in the London basin by high-pressure skies above have attracted many different monikers, including 'London fog', 'smog', 'pea souper' and 'London particular'.

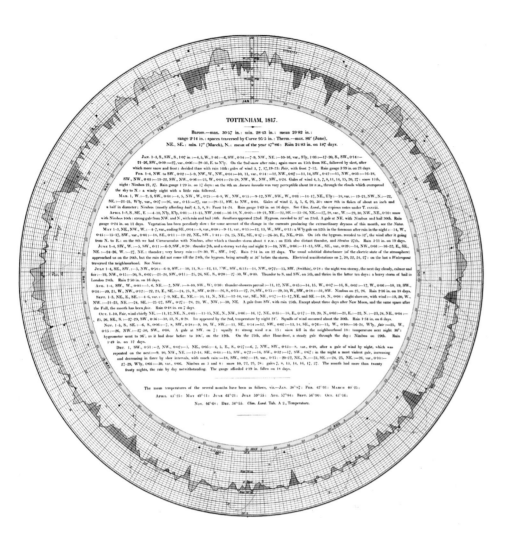

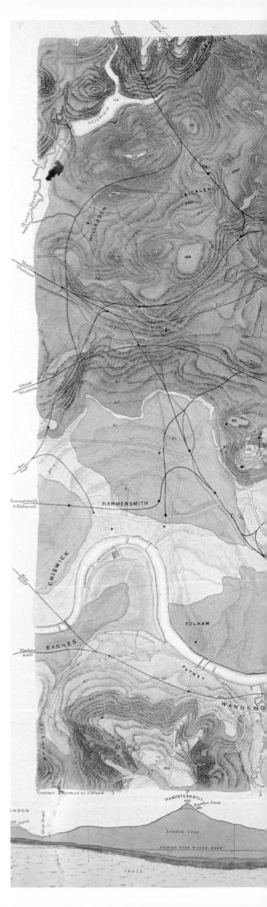

Autographic Curve, Tottenham, 1817
Luke Howard

Luke Howard—the great nineteenth century amateur meteorologist who developed the nomenclature of cloud systems—was also devoted to the weather and climate of London. He spent a lifetime recording and describing the daily climate and used his data to map London in ways that had never been attempted before. His autographic curve describes a year's weather patterns in the area of Tottenham (where he resided) and sets a high standard for the graphic clarity of London data.

Image courtesy of Science & Society Picture Library.

Geological Map of London, 1871
Robert William Mylne

Robert Mylne's geological map of the Thames basin follows on from William Smith's pioneering map of the whole of England published in 1815. This map, however, was developed specifically for water companies to satisfy their interest in the geological strata underneath London's streets. As with Smith's map, contour lines are employed to describe what cannot be directly seen.

Image courtesy of the Guildhall Library, City of London.

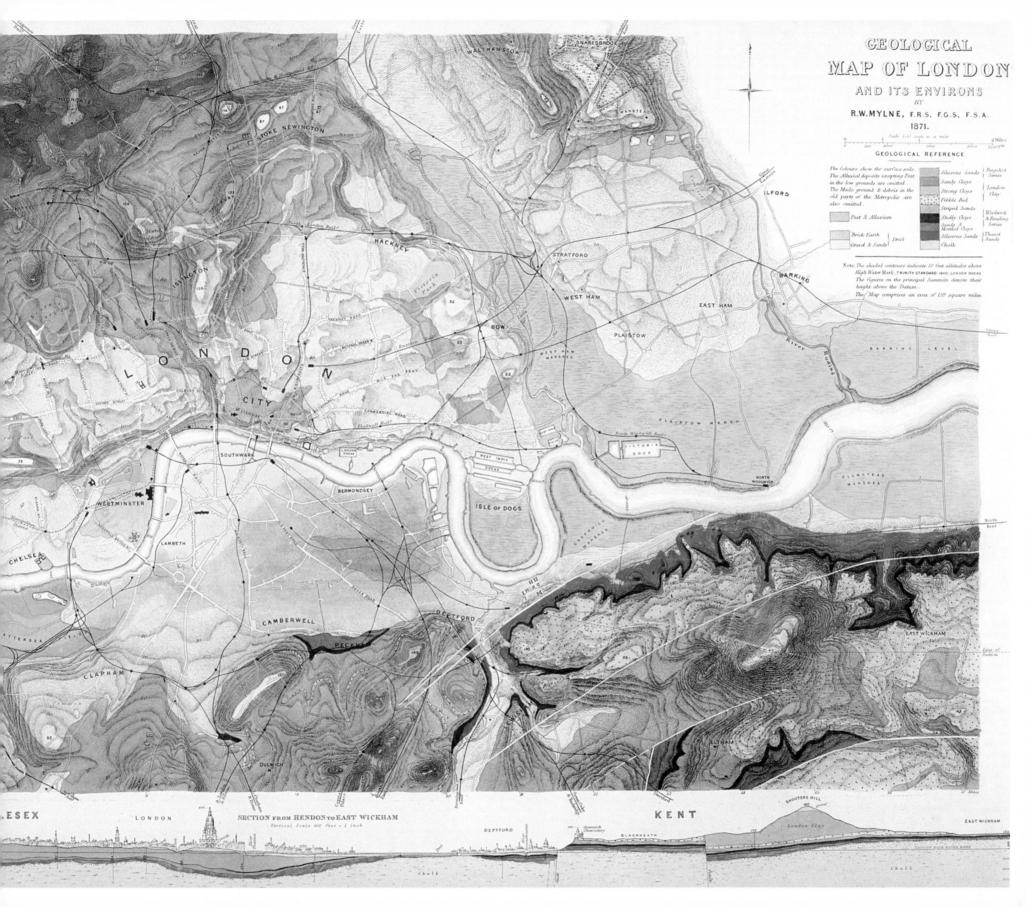

GEOLOGICAL
MAP OF LONDON
AND ITS ENVIRONS
BY
R.W.MYLNE, F.R.S. F.G.S. F.S.A.
1871.

GEOLOGICAL REFERENCE

The Colours show the surface soils.
The Alluvial deposits excepting Peat
in the low grounds are omitted.
The Made ground & debris in the
old parts of the Metropolis are
also omitted.

Siliceous Sands	Bagshot Series
Sandy Clays	London Clay
Strong Clays	
Pebble Bed	
Striped Sands	Woolwich & Reading Series
Shelly Clays	
Sandy & Mottled Clays	
Siliceous Sands	Thanet Sands
Chalk	

Peat & Alluvium

Brick Earth
Gravel & Sands Drift

Note. The shaded contours indicate 10 feet altitudes above
High Water Mark. TRINITY STANDARD. 1800. LONDON DOCKS.
The figures on the principal Summits denote their
height above the Datum.
The Map comprises an area of 159 square miles.

SECTION FROM HENDON TO EAST WICKHAM
Vertical Scale 500 feet = 1 inch

ESSEX   LONDON   KENT   SHOOTERS HILL   EAST WICKHAM
DEPTFORD   Greenwich Observatory   BLACKHEATH   London Clay
Chalk   Chalk

WALTHAMSTOW   SNARESBROOK   WANSTEAD
STOKE NEWINGTON   ILFORD
HIGHGATE   HACKNEY   STRATFORD   BARKING
ISLINGTON   WEST HAM   EAST HAM
LONDON   BOW   PLAISTOW   BARKING LEVEL
CITY   VICTORIA DOCK   PLAISTOW MARSH
SOUTHWARK   LONDON DOCKS   WEST HAM MARSHES
WESTMINSTER   BERMONDSEY   WEST INDIA DOCKS   NORTH WOOLWICH   PLUMSTEAD MARSHES
LAMBETH   ISLE OF DOGS   GREENWICH MARSHES   NORTH KENT
CHELSEA   EAST WICKHAM
BATTERSEA FIELDS   DEPTFORD   ELTHAM
CAMBERWELL   Line of Section
CLAPHAM   PECKHAM
DULWICH

They are now, as they were then, part of the popular romantic image of old London, and feature strongly in Victorian literature as well as its mythology and exaggerated fear of crime. If smog didn't have a significant impact on crime, it was deadly in other ways, with the Great Smog of 1952 reportedly killing over 4,000 people across the city and asphyxiating a herd of cattle at Smithfield market.

The Great Smog led directly to the Clean Air Acts of 1954, 1956 and 1968. Over time, these acts banned the emission of black smoke and required conversion to smokeless fuel. Except for two further smogs in 1957 and 1962, London has been relatively clean—and much brighter —ever since. However, despite the heavy sooty and acidic atmosphere no longer being a threat, London's air is now polluted with a wide range of other chemicals, gases and particulates and increased monitoring (and action) has become a necessity.

In response, mapping London's climate and pollution levels has become a growing sector. Maps are available showing levels of carbon monoxide, nitrogen dioxide, ground level ozone, particulate matter, sulphur dioxide, hydrocarbons and lead; both as historical data and up-to-date forecasts. Similar information is available on noise levels and flood risks, in addition to the normal day-to-day weather information concerning temperature, rainfall and humidity. Other more creative approaches to showing the quality of the environment include maps showing ladybird densities across London and arousal maps based on the skin measurements recorded as pedestrians move around the city.

As the environment and climate change become key policy arenas and important aspects of everyone's lives, good information and accurate predictions will be needed to maintain lives and lifestyles. Precise and tailored mapping is required to provide real-time information on pollution, heat density and the most energy efficient means of keeping buildings comfortable and usable. The impact of rising sea levels, flash floods, droughts and prolonged heat waves must be planned and prepared for, while methods for assessing the effect of building development on microclimates should ensure that potential damage is mitigated. As a result, the next ten years are likely to see environmental mapping develop so that they become as useful as street atlases are today.

Land Surface Temperature Variations, 2004
King's College, London

This image depicts data from the ASTER 90 metre
spatial resolution satellite, which has been processed
to show daytime land surface temperature variations
across London on the morning of 9 September 2004.

Image courtesy of Professor Martin Wooster and Dr
Weidong Xu, Environmental Monitoring and Modelling
Group, Department of Geography, King's College
London. ASTER data are courtesy of NASA (USA) and
METI/ERSDAC (Japan).

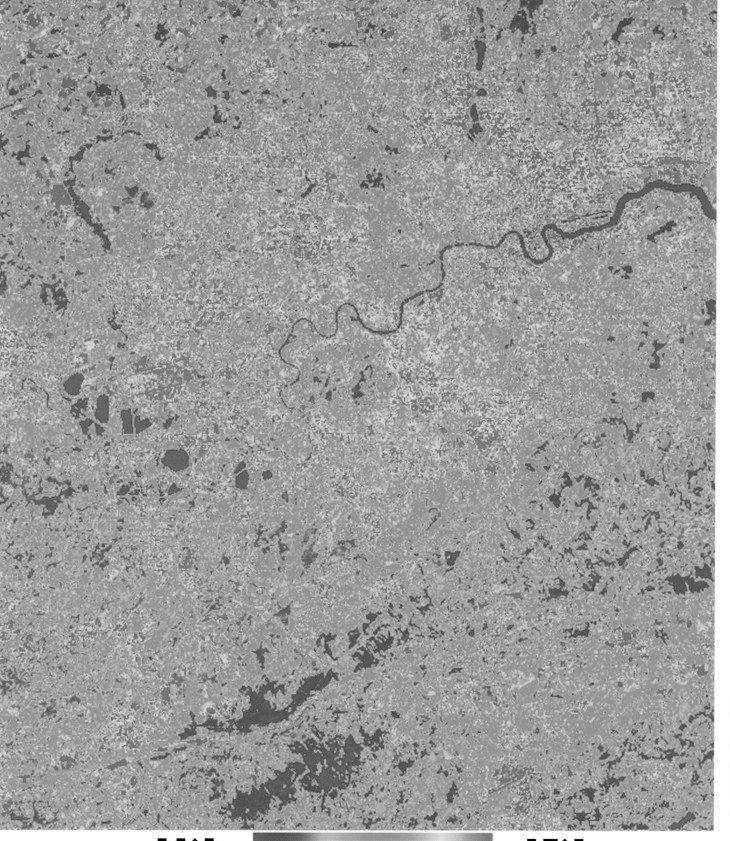

20°C 35°C

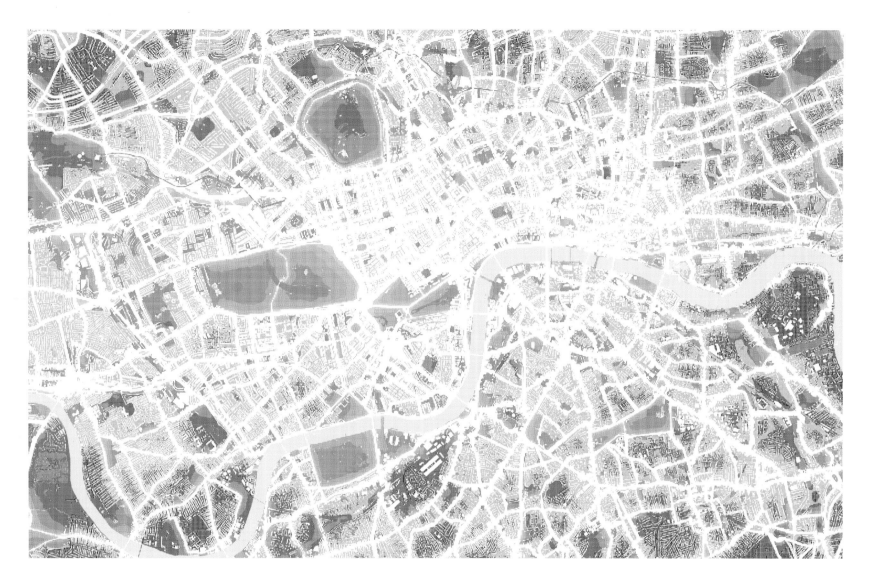

Silent London, 2005
Simon Elvins

Simon Elvins has used information collected by the
government—concerning the noise levels within
London—to plot a map of the capital's most silent
spaces. The map is intended to reveal a hidden
landscape of quiet spaces and show an alternate side
of the city that would normally go unnoticed.

Image courtesy of Simon Elvins.

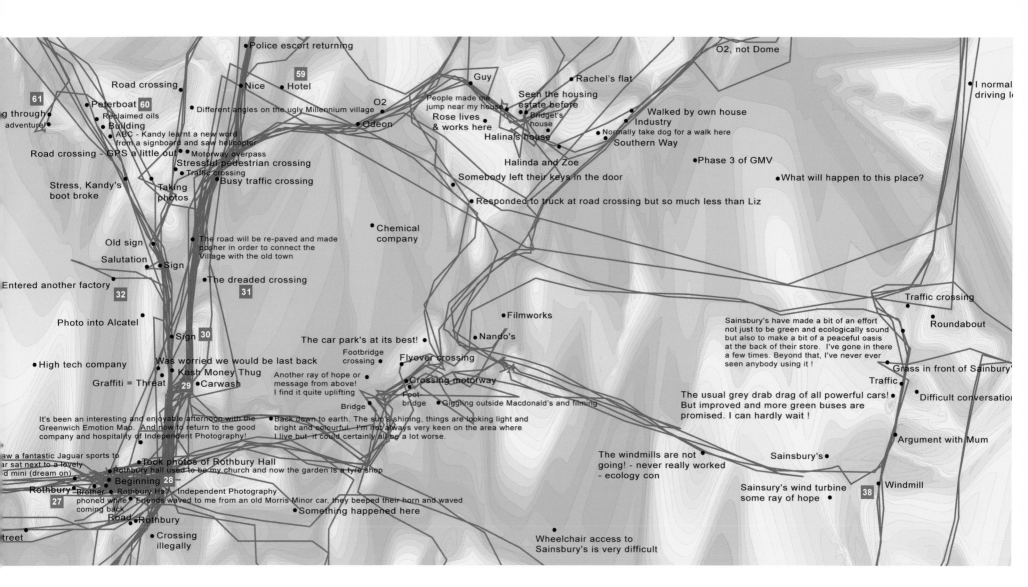

I normal[ly]
driving [le]

O2, not Dome

Police escort returning

Guy • Rachel's flat

Road crossing   Nice • Hotel   59

61   Seen the housing
                                      estate before
                          O2   • Industry • Walked by own house
g through   Peterboat 60   People made me   Bridget's
adventure!   Reclaimed oils jump near my house house
        Building   Odeon Rose lives
        ABC - Kandy learnt a new word & works here   Normally take dog for a walk here
Road crossing - GPS a little out from a signboard and saw helicopter Halina's house Southern Way
        • Motorway overpass
        Stressful pedestrian crossing Halinda and Zoe   • Phase 3 of GMV
Stress, Kandy's Traffic crossing
boot broke Taking Busy traffic crossing Somebody left their keys in the door • What will happen to this place?
        photos
                              • Responded to truck at road crossing but so much less than Liz

Old sign   The road will be re-paved and made
Salutation   posher in order to connect the • Chemical
        • Sign   Village with the old town   company

Entered another factory • The dreaded crossing
        32   31
                                                              • Traffic crossing
Photo into Alcatel • Sign 30   • Filmworks Sainsbury's have made a bit of an effort • Roundabout
                                                             not just to be green and ecologically sound
• High tech company   Was worried we would be last back • Nando's but also to make a bit of a peaceful oasis
        • Kash Money Thug The car park's at its best! at the back of their store. I've gone in there
        29   • Carwash Footbridge   a few times. Beyond that, I've never ever • Grass in front of Sainsbury['s]
Graffiti = Threat   crossing • Flyover crossing seen anybody using it !
                    Another ray of hope or   • Traffic
                    message from above!   • Crossing motorway   • Difficult conversation[s]
                    I find it quite uplifting Foot-
It's been an interesting and enjoyable afternoon with the bridge • Giggling outside Macdonald's and filming The usual grey drab drag of all powerful cars!
Greenwich Emotion Map. And now to return to the good • Bridge   But improved and more green buses are
company and hospitality of Independent Photography! • Back down to earth. The sun's shining, things are looking light and promised. I can hardly wait !
        bright and colourful. I'm not always very keen on the area where
[s]aw a fantastic Jaguar sports to I live but it could certainly all be a lot worse.   • Argument with Mum
[ca]r sat next to a lovely Took photos of Rothbury Hall
[ol]d mini (dream on) Rothbury hall used to be my church and now the garden is a tyre shop The windmills are not Sainsbury's •
Rothbury Beginning 28 going! - never really worked
27 Brother Rothbury Hall Independent Photography - ecology con
        phoned while Friends waved to me from an old Morris Minor car, they beeped their horn and waved Sainsury's wind turbine 38 • Windmill
        coming back • Something happened here some ray of hope
        Road Rothbury
[s]treet • Crossing • Wheelchair access to
        illegally Sainsbury's is very difficult

Greenwich Emotion Map, 2005–2006

Christian Nold

A map of Greenwich derived from the responses
of volunteers adorned with galvanic skin monitors
(devices that measure the small changes in skin pore
size and sweat gland activity) who are tracked using
global satellite positioning devices. Through this
playful use of available technology, a recording of the
emotions evoked by the Greenwich townscape can be
read in map form.

Image courtesy of Christian Nold..

SERVING
THE CITY

BOUNDARIES—POLITICAL ORGANISATION

Since its foundation, London has been divided into distinct geographical entities—the City, Westminster, the Liberties, wards, parishes, boroughs and councils—and for centuries different laws, taxes and services applied, depending on the jurisdiction one found oneself in. For everyone concerned, whether administrators, aldermen, the police or the general public, it was important to know where one was and where the administrative boundaries lay. Maps were and are, of course, ideal for making such distinctions; boundaries can be delineated and areas elaborately colour-coded. However because drawing such boundaries was so easy, this resulted in a multilayered and deeply confusing patchwork of zones and districts carved up into a myriad of separate areas.

Elaborate divisional boundaries exist for the various services that are necessary for London to function. Only very rarely have they had congruent boundaries. Such services, whether the police, the post, drainage, water supply, transport, fire, health trusts, and so on, have developed their own historical divisions and sub-divisions. Tony Travers, writing about the Victorian metropolis, commented: "There was no uniformity and no requirement for consistent provision. Local and general acts randomly provided the capital with the services necessary for such a large agglomeration of people."[1] Such relative administrative chaos and complexity has inevitably relied on maps to help gain some degree of control and understanding, and mapmakers have certainly not shied away from creating yet more lines to separate one authorities, boards or companies fiefdom from another.

A few boundaries reflect logical divisions in London's geography—a local administrative area will very rarely cross the Thames, for example—but most have grown from a combination of historical accident and frequently from the pursuit of political or commercial advantage. The result is London's haphazard form; its powerful rivalries, relatively impotent governance, lack of any obvious centre and its idiosyncratic and conflicting street patterns. Until the mid-twentieth century, and as the metropolis grew, new conglomerations and developing areas would adopt and adapt existing village administrations and invent new arrangements, services and administration—further adding to the complexity and chaos.

Political boundaries first appear after the Great Fire on the Exact Surveigh of the Ruines of London, in 1667—a map that was prepared by a team (that included Hollar) directed by surveyor John Leake.

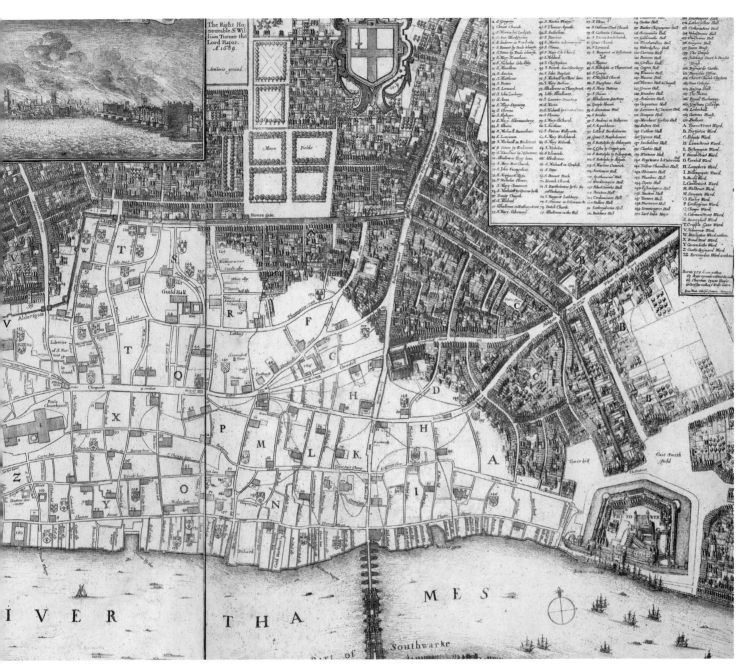

An exact Surveigh of the Streets, Lanes and Chvrches contained within the Rvines of the City of London first described in six plats, 1667
John Leake

The Great Fire of London in 1666 forced the authorities into unprecedented administrative action. Any records they had of parish boundaries, buildings and land ownership was scanty but, nonetheless, essential to any rebuilding effort. Wenceslaus Hollar had rapidly produced a plan of the destroyed area, but a more systematic effort was required. A survey team was assembled under the leadership of John Leake and within the year they produced this accurate assessment of the burnt out area. The survey map was engraved by Hollar, who re-used his earlier image of the houses and streets beyond the survey zone—still in the earlier bird's-eye style—as well as an image of the burning city.

This map is the first to show the individual parishes fitting together and their freeform boundaries crossing plots of land. The map facilitated the resolution of land ownership issues in the city and rebuilding could start more quickly than anyone had anticipated. As a result, London was back in business.

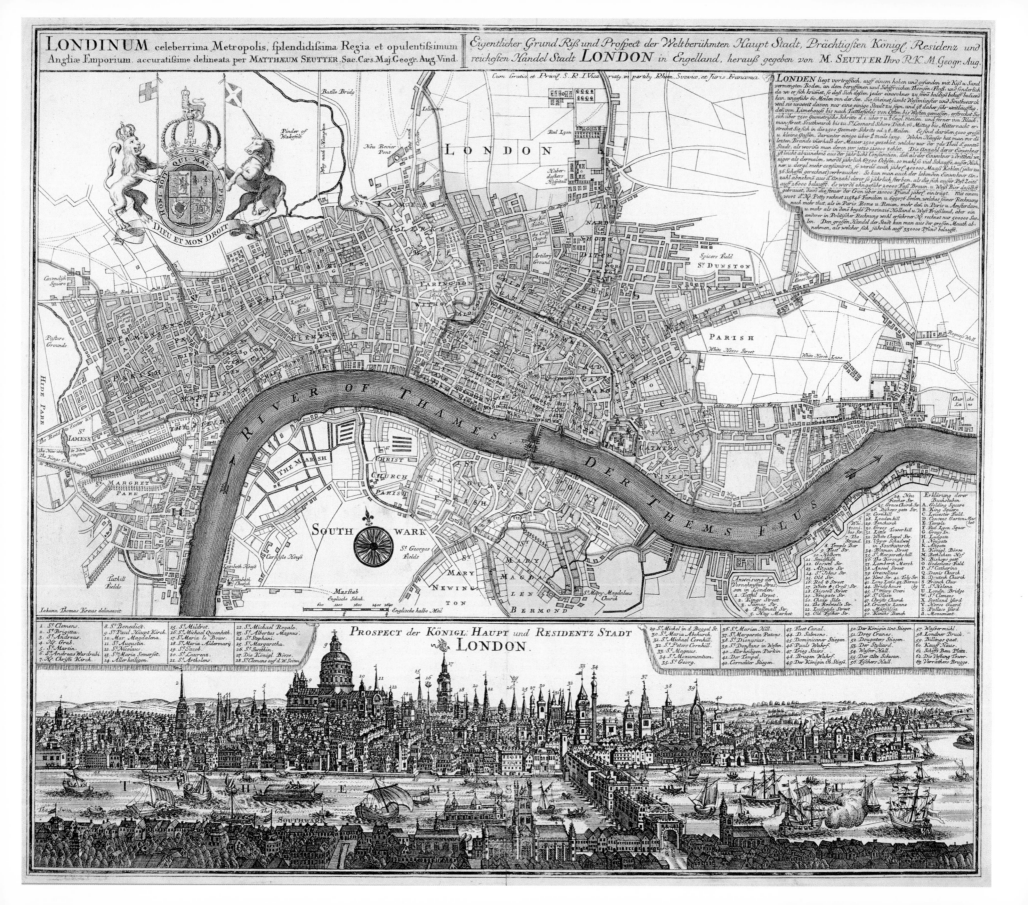

Map of Westminster, the City of London
and Southwark, 1720
Matthaeus Seutter

A hand coloured map carefully organised to show the
arrangement of the wards and parishes of London.
The administrative and governance arrangements of
London were the responsibility of these sub-regions
until the Victorians instituted modest reforms to
run infrastructure and services over 100 years later.
London's compartmentalised governance is responsible
for much of the city's character and variety as well as its
general lack of early citywide initiatives. In many ways,
the image of the innumerable competing church

spires at the bottom of the map sums up the spirit of
administrative organisation in London with parishes,
wards, paving commissions, private service companies,
and many others, all ruling their individual fiefdoms.

Image courtesy of the Guildhall Library, City of London.

A number of dotted lines show the boundaries between the city wards in
the fire-destroyed areas. No such borders appear in Hollar's own similar
map of the previous year. Ten years later, in Ogilby and Morgan's Large
and Accurate Map of the City of London, 1676, these dots have become
streams of tiny bubbles that show the boundaries of ward and parishes
and reveal the minute and irregular shapes of the streets and houses
included in each one. Morgan repeats this style of marking border lines in
his later maps, as they extend to wider areas around London—a method
which later became the standard approach adopted by numerous other
mapmakers. The dotted lines, in turn, became essential guides to the army
of hand colourists individually embellishing and enriching the maps.

For most of its history, outside the confines of the City, London
relied on governance by the vestries of the parishes, precincts and liberties
(both open: elected by ratepayers, or closed and select: appointed by
a hereditary group of about 30 'principal inhabitants'). These bodies
overlapped with further boards and commissions responsible for paving,
public health lighting, cleansing and sewers. The autonomy and self-
governing nature of these bodies was seen as a matter of political and
moral principle and part of London's ancient character and freedom.

As a result, the best maps of the early eighteenth century are those
of individual parishes and wards, such as those of Richard Blome
published by John Strype in his much expanded version of Stow's
Survey of London, 1720, and by Benjamin Cole, published in William
Maitland's History of London, 1756. Although such maps barely
need boundary markers, as they are not much concerned with the
world beyond their borders, they are still there, running along the
outer edges of the mapped area. The parishes depicted in these maps
were commonly an administrative force unto themselves and were
maintained that way for centuries. When (relatively modest) city-
wide administration was finally forced on them by the drainage and
public health crisis that culminated in the Great Stink of 1858, it
was fought as an affront "to our principles and our taste, which have
hitherto always encouraged local self-government".[2]

Above the level of the wards and parishes lay the four-fold division
of London into the City, Westminster, Southwark and the surrounding
County of Middlesex. Each of these returned two MPs to Parliament. These
eight members represented over 13 per cent of the population but hardly
counted among the 505 MPs representing the rest of England and Wales.

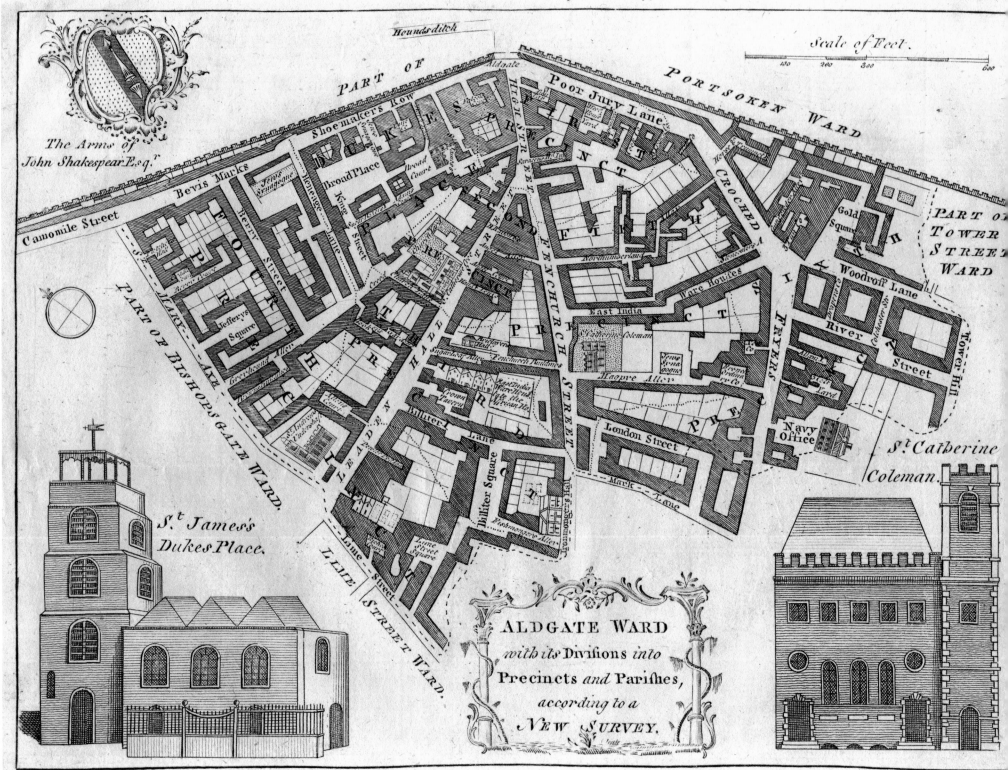

Scale of Feet.

The Arms of
John Shakespear Esq.r

Canomile Street

PART OF

Houndsditch

Bevis Marks

PORTSOKEN WARD

Poor Jury Lane

PART OF
TOWER
STREET
WARD

St James's
Dukes Place.

ALDGATE WARD
with its Divisions into
Precincts and Parishes,
according to a
NEW SURVEY.

St. Catherine
Coleman.

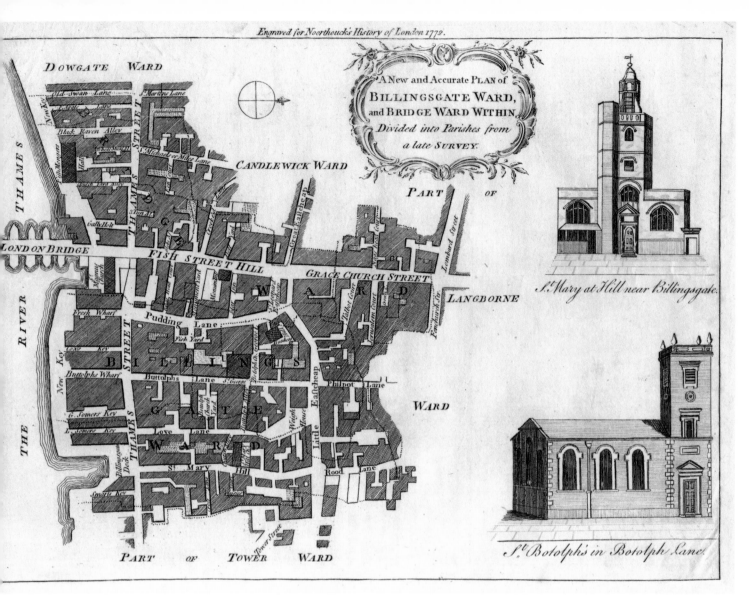

A New and Accurate PLAN of BILLINGSGATE WARD, and BRIDGE WARD WITHIN Divided into Parishes from a late SURVEY.

St Mary at Hill near Billingsgate.

St Botolph's in Botolph Lane.

Opposite, right and following pages:

Aldgate Ward with its divisions, Billingsgate Ward and Bridge Ward Within, Bishops Gate Ward Within and Without and Farringdon Ward Without, 1772 John Noorthouck

As the effective authorities, the parishes and City Wards required accurate maps of their domains for both administrative and decorative purposes. So intricate were their shapes, that it was difficult to provide them with a definite identity. In 1720 John Strype, along with Richard Blome, published a full set of parish maps to accompany his new edition of Stow's *Survey of London*.

Aldgate, Billingsgate and Bishops Gate (now Bishopsgate) are inner city wards while Farringdon Without, as its name suggests, lies outside the city walls. Having developed later, as the city spread westwards towards Westminster, Farringdon Without (there was also a part of the Ward Within) had far more room to expand and became much larger than the inner city wards.

Images courtesy of David Hale/MAPCO.

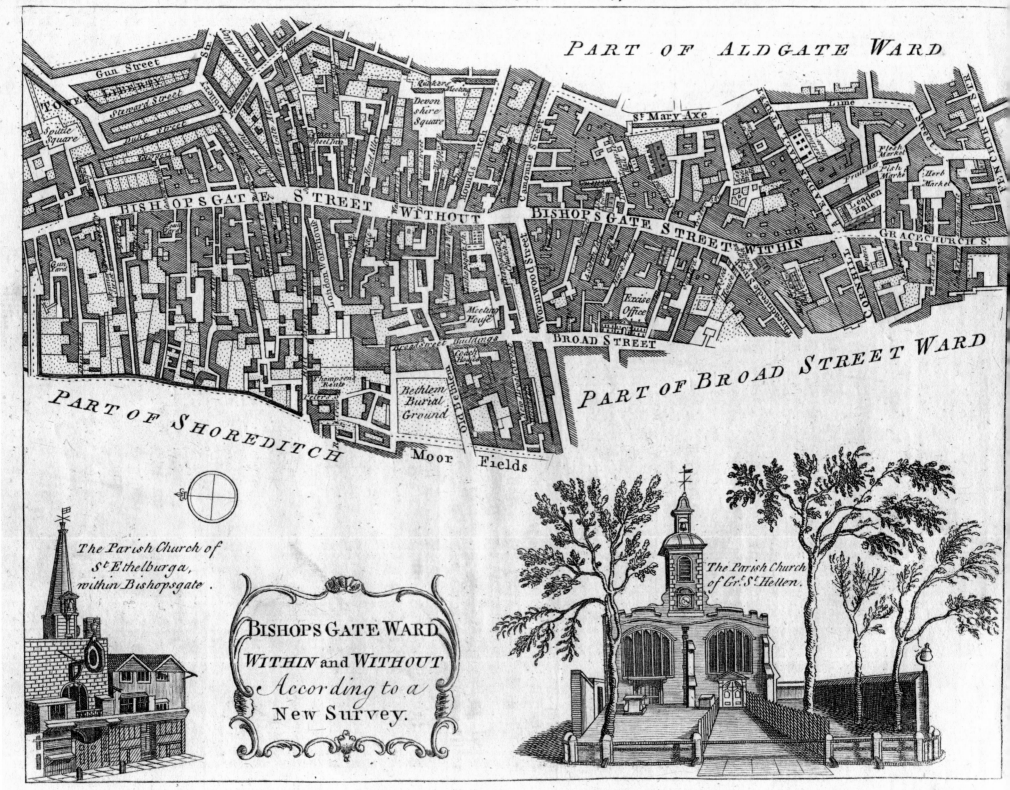

PART OF ALDGATE WARD

Gun Street

TOWER LIBERTY

Stewart Street

Duke Street

Spittle Square

Devonshire Square

Quakers Meeting

St Mary Axe

Lime

Flesh Market

Fish Market

Herb Market

Leaden Hall

BISHOPSGATE STREET WITHOUT

BISHOPSGATE STREET WITHIN

GRACECHURCH S.

Gun Yard

Rounds Ditch

Camomile Street

Wormwood Street

Meeting House

Excise Office

BROAD STREET

Broad Street Buildings

Coach House

Thompson's Rents

Bethlem Burial Ground

PART OF SHOREDITCH

Moor Fields

PART OF BROAD STREET WARD

The Parish Church of St Ethelburga, within Bishopsgate.

BISHOPSGATE WARD WITHIN and WITHOUT According to a New Survey.

The Parish Church of Gt St Hellen.

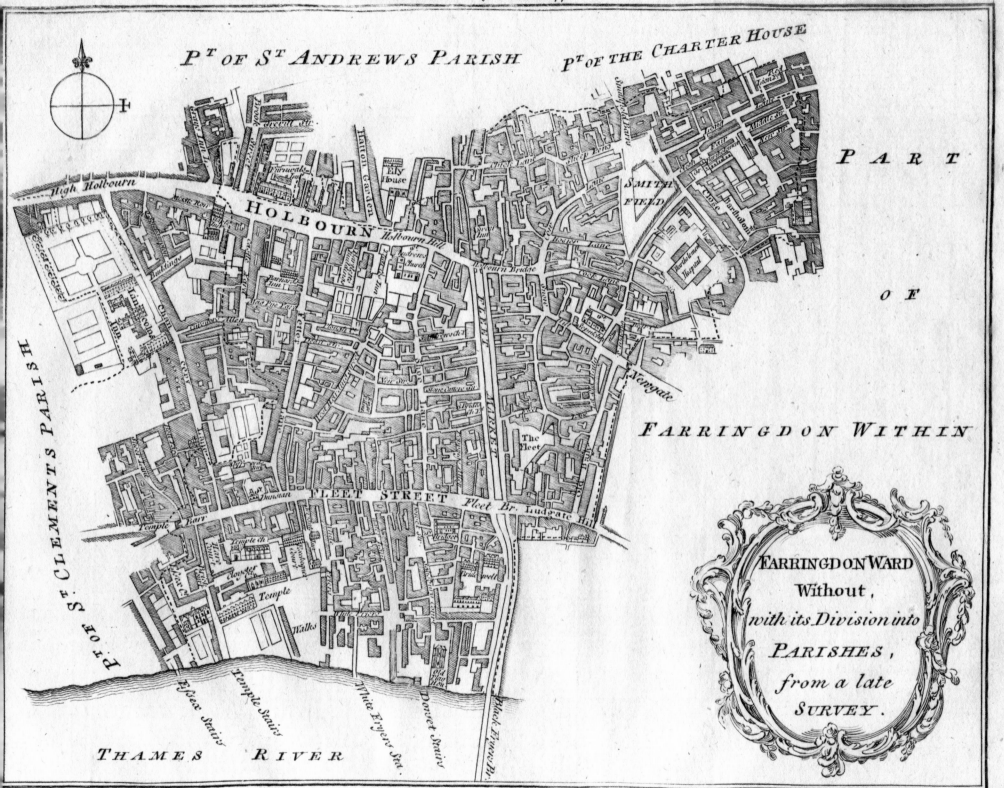

Pt OF St ANDREWS PARISH

Pt OF THE CHARTER HOUSE

PART

OF

FARRINGDON WITHIN

Pt OF St CLEMENTS PARISH

THAMES RIVER

FARRINGDON WARD Without, with its Division into PARISHES, from a late SURVEY.

The 1832 Reform Act redefined the boundaries of London and created five new boroughs (Marylebone, Finsbury, Tower Hamlets, Greenwich and Lambeth) increasing the number of MPs to 22. So began the regular practice of redrafting electoral boundaries and the necessary creation of new maps to explain them. (Although London had to wait through the Second Reform Act of 1867, which created the boroughs of Hackney and Chelsea, for the 1885 Redistribution of Seats Act, to elect a number of MPs proportional to its population.)

Governing and serving London has never been easy, a role compounded by the city's dominant position in the UK's economy. Counter to the parochial view and isolationism, there has been a necessary—though almost always politically thwarted—struggle to unite its government and administration into a single meaningful whole. The compromises resulting from such efforts have instead succeeded in adding more layers of complexity to London's administrative maps.

The first pan-London administrative bodies; made necessary by the lack of a London government, were the Metropolitan Police, established in 1829, the Metropolitan Board of Works (MBW), established in 1855 and the London School Board in 1870. The vestries and their associated commissions and boards, having avoided abolition due to the partial reforms of 1855, nonetheless continued to govern locally in their own highly individual ways, and in 1885 *The Times* commented: "Within the Metropolitan Limits the local administration is carried on by no fewer than 300 different bodies, deriving powers from about 250 local Acts."[3]

The arrival of the MBW with its wide-ranging responsibilities—if few resources—for drains, main roads and bridges, parks and the administration of the 1855 Building Act, as well as the naming of streets and numbering of houses, meant that new, accurate and definitive administrative maps of London were required. The most recent comprehensive survey and map, by the Greenwoods in 1827, was, by then, very out-of-date. Although the newly established Ordnance Survey had carried out a survey for the Metropolitan Commissioners for Sewers in 1848, it was only drawn in skeleton form and contained very little detail. In March 1862, the MWB determined: "That a map of the whole area within the jurisdiction of the Board be provided, defining the boundaries of the various vestries and district boards. The Works and Improvements Committee were instructed to consider and report upon the best means of providing such a map."[4]

The Metropolitan Boroughs as defined by the
Reform Bill, 18132
Robert Dawson

A map produced for the 1832 Reform Act, which created
five new London boroughs and modestly increased the
number of MPs representing the boroughs to 22. The Act
also extended the vote to men over the age of 21 (who
owned or rented property "of clear legal value of not
less than ten pounds") but specifically disenfranchised
women. The passing of the various Reform Bills took
three attempts, during which time governments were
brought down and numerous riots sparked.

The boroughs created acted as electoral districts
until the reforms of the London Government Act,
1899, which again divided the County of London into
28 metropolitan boroughs (not including the City of
London), this time with administrative powers. They
replaced the 41 Parish vestries and District Boards of
Work as the local tier of London Government.

Image courtesy of the Guildhall Library, City of London.

Stanford's Library Map of
London and its Suburbs, 1862
Edward Stanford

With the formation of the MBW in 1855, along with other pan-London administrative bodies, maps were needed to represent their areas of responsibility. Edward Stanford published the Library Map in 1862, providing London government with the mapping resources it needed to operate effectively. Stanford continued to publish versions of the Library Map in 21 editions up to 1901.

The map provided an authoritative view of a wide area, both in and around London, partly coloured to support many different uses and readings. This information was also used by Stanford to provide a range of specifically administrative maps. They were well-designed, attractive and functional but unashamedly serious and invariably aimed at purchasers who could afford such large-scale publications.

Image courtesy of the Guildhall Library, City of London.

The MWB failed to find a means of providing such a map. Instead, the task was undertaken by the emerging mapmaker Edward Stanford, who, seeing an opportunity, decided to expand on the series of Library Maps of the world he had just embarked upon, by including detailed maps of London. The large-scale New Map of London was launched and advertised in *The Times* on 28 May 1862 with an explicit boast of "its adaptability for various administrative purposes" and it rapidly became the *de facto* administrative map for the capital. The map, in its various forms, was used as the base for a wide range of over-printing that showed boundaries for municipal, ecclesiastical, electoral, health and service utility purposes (many specifically for the MWB), as well as being used to show transport routes and proposals and social and statistical information, including Charles Booth's poverty maps. Stanford's Library Map went through a continuous series of new editions and updating until at least 1901 and may have effectively kept the Ordnance Survey out of the picture until relatively late in its survey and mapping of Britain.

The London County Council (LCC) was created in 1888 to fill the long-standing administrative and democratic vacuum, with the first elections held for it in 1899. In addition to the LCC's other range of powers and responsibilities it took over the activities of the MBW and later absorbed the London School Board in 1904. The lower tiers of London government were reformed in 1899 with the amalgamation of the vestries and parishes into metropolitan boroughs. Stanford immediately published a map to show the new boundaries—Stanford's Map of the County of London—that complied with the Local Government Act of 1988, by showing the urban sanitary districts as well as the new boroughs and electoral areas. At the end of the nineteenth century, the newly published Ordnance Survey maps of London provided the LCC with the opportunity to produce their own maps. From this time onward, political and administrative maps tended to be produced by the public authorities concerned, or be drawn over Ordnance Survey base-maps; no doubt to Stanford's considerable disadvantage.

Despite new government arrangements, the developing areas of London continued to proliferate, and the area covered by the LCC simply became a core of the greater metropolis. By 1939, city development covered more than twice the land area it had in 1914. Other administrative organisations, including the London Passenger Transport Board (London Transport), established in 1933, covered a much larger portion.

A New Map of the Country in the vicinity of
London—Metropolitan Boroughs and Parish
Boundaries, 1872
James Wyld

Almost all publishers of London maps provided
information about the city's confusing political
boundaries. This map published by James Wyld, like his
father of the same name (the Geographer Royal), has
been coloured along the county boundaries, such as
Middlesex, Essex, Kent and Surrey, surrounding London.

Image courtesy of David Hale/MAPCO.

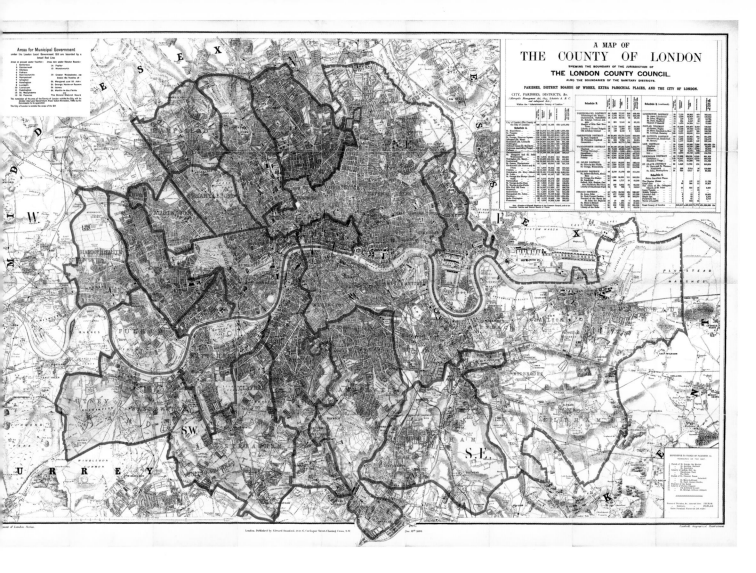

A Map of the County of London
(the London County Council), 1898
Edward Stanford

The LCC was created in 1889 by the Local Government
Act 1888. This map, created in conjunction with the
London Government Act just ten years later, shows the
area of jurisdiction of the LCC as well as the Sanitory
Districts (where the newly created boroughs are
outlined in red and coloured in). The inset box provides
a series of statistics covering vestries, areas, inhabited
houses, population and rateable value; indicating that
the map was, most probably, intended for the walls of
bureaucrat offices.

This was a time of administrative turmoil for
London government, as the 38 vestries were
dismantled and their powers passed over to the 28
boroughs. Some of these were direct transfers that
covered the same administrative areas but others
were amalgamations and mergers.

Image courtesy of David Hale/MAPCO.

Following pages:

Ecclesiastical Boundaries, 1877
Edward Stanford

The parish boundaries shown on this map, in which
the complex City of London is wisely ignored, were the
administrative districts of the wider city at this time.
The internal politics of the parish vestries endlessly
frustrated much real reform. A map, such as this one
published by Stanford, would have been essential to
anyone attempting to provide services in London or
who had to deal with the bureaucracy needed to obtain
consents and licences.

Image courtesy of David Hale/MAPCO.

HAMPSTEAD

TRINITY
WEST HAMPSTEAD
1,000

CHRIST CHURCH
1,094

ST MARY KILBURN
4,299

HOLY TRINITY KILBURN
8,000

ST LUKE KILBURN

ST AUGUSTINE KILBURN
4,846

ST JOHN THE EVANGELIST KILBURN
5,000

ST JOHN KENSAL GREEN

KENSAL GREEN
5,387

ST PETER PADDINGTON
7,161

ST ANDREW AND ST PHILIP NOTTING HILL
5,523

ST MARK
5,689

ST MARK NOTTING HILL
18,000

ALL SAINTS PADDINGTON
15,221

NOTTING HILL

ST CLEMENT KENSINGTON

ST JOHN NOTTING HILL
6,860

ST JAMES NORLAND SQ
6,514
10,000

ST PETER NOTTING HILL

ST GEORGE CAMDEN HILL
8,200

ST MATTHEW BAYSWATER

ST MATTHEW HAMMERSMITH
5,000

ST MARY ABBOT

KENSINGTON
14,830

ST BARNABAS

ST PHILIP EARLS COURT

ST PETER BELSIZE PARK
2,767

ST PAUL AVENUE ROAD
3,071

ALL SOULS HAMPSTEAD
2,608

ST MARY THE VIRGIN PRIMROSE HILL & PARK

ST SAVIOUR SOUTH HAMPSTEAD
5,663

9,807

ST MARK REGENT'S PARK

ALL SAINTS ST JOHNS WOOD
5,268

ST STEPHEN THE MARTYR
9,836

ST MARK HAMILTON TERRACE
5,054

ST SAVIOUR PADDINGTON
6,816

EMMANUEL
9,000

ST MARY MAGDALEN PADDINGTON
9,900

ST PAUL PADDINGTON
5,000

ST MATTHEW
7,660

ST PAUL LISSON GROVE
6,793

CHRIST CHURCH
11,467

ST MARY
10,719

ST BARNABAS
6,356

ST MICHAEL PADDINGTON
5,350

ST MARK
4,250

ST MARY BRYANSTON SQUARE
19,301

ST LUKE
7,392

ALL SAINTS PADDINGTON
5,488

ST JOHN PADDINGTON
6,030

ST JAMES PADDINGTON
6,110

CHRIST CHURCH

ST STEPHEN PADDINGTON
6,600

HOLY TRINITY PADDINGTON
13,685

LANCASTER GATE

6,790
5,700

MARYLEBONE
22,228

ST THOMAS
9,790

TRINITY MARYLEBONE
14,020

ST JOHN FITZROY SQUARE
4,260

ALL SOULS
14,811

ST MARY MAGDALENE MUNSTER SQ
5,489

ALL SAINTS ST PANCRAS
6,018

ST SAVIOUR

ST PANCRAS
17,201

ST PANCRAS ROAD

HOLY TRINITY HAVERSTOCK HILL

ST ANDREW HAVERSTOCK HILL

ST LUKE HOLLOWAY
6,508

ST LUKE KENTISH TOWN
8,118

HOLY TRINITY

ST PAUL CAMDEN SQ
3,865

8,822
15,516
13,922

ST THOMAS AGAR TOWN
7,338

ST MICHAEL ISLINGTON

ST STEPHEN
6,006

ALL SAINTS ISLINGTON
14,517

CAMDEN TOWN
14,011

ST MATTHEW OAKLEY SQ
8,458

OLD ST PANCRAS
7,930

CHRIST CHURCH ALBANY STREET
10,010

ST MARY SOMERS TOWN
8,643

CHRIST CHURCH PENTONVILLE
8,028

ST MARY SOMERS TOWN
9,852

ST JAMES HAMPSTEAD ROAD
5,604

ST PETER REGENT SQUARE

HOLY CROSS
6,000

ST BARTHOLOMEW

ST GEORGE THE MARTYR
17,692

ST JOHN HOLBORN
4,700

ST GILES IN THE FIELDS
35,703

ST ANDREW WELLS STREET
3,044

ALL SAINTS

ST ANDREW

ST LUKE
11,605

ST PAUL COVENT GARDEN
4,376

ST CLEMENT

ST ANNE SOHO
8,000

HANOVER CHAPEL
5,501

ST MARY AUDLEY ST
4,287

ST THOMAS REGENT ST

ST PETER WINDMILL

ST GEORGE HANOVER SQUARE
15,530

ST JAMES PICCADILLY WESTMINSTER
11,472

CHRIST CHURCH MAY FAIR
6,287

BERKELEY CHAPEL

HYDE PARK

REGENT'S PARK

KENSINGTON GARDENS

HOLY TRINITY BROMPTON

ST STEPHEN KENSINGTON
2,322

ALL SAINTS KNIGHTSBRIDGE
6,266

HOLY TRINITY KNIGHTSBRIDGE

SAVIOUR CHELSEA
8,766

HOLY TRINITY CHELSEA
11,268

ST PETER PIMLICO

ST ANDREW WESTMINSTER
4,063

CHRIST CHURCH BROADWAY

ST MARGARET WESTMINSTER
4,757

WESTMINSTER
15,075

ST MARTIN IN THE FIELDS

ST MARY

4,713

CITY of LONDON.

WESTMINSTER.

School Board Maps of London
—City of London and Westminster, 1877
Edward Stanford

Local school boards were established by the 1870 Education Act and, as a result of amendment proposed during its passage by the MP for Finsbury, William Torrens, London had a single directly elected board administered by the same overall area (covering 117 square miles in total) as the MBW. The electoral divisions for the Board elections were based on London's borough constituencies of the City, Southwark, Chelsea, Greenwich, Lambeth, Tower Hamlets, Hackney, Westminster, Finsbury and Marylebone.

The maps were intended solely for administrative purposes: for identification of school locations, gaps in provision and the planning of new schools. They were not meant to be distributed widely, as the Board only ordered 200 copies from Stanford for the direct use of its members, supervisors, inspectors and administrators.

Images courtesy of David Hale/MAPCO.

Following pages:

County of London Plan,
Social and Functionnal Analysis, 1943
JH Forshaw and Patrick Abercrombie

During the Second World War, two monumental studies were undertaken on the future of London; *The County of London Plan*, 1943, and the *Greater London Plan* published a year later. The first of the plans, covering the LCC administrative area, clearly showed that London mapping needed to encompass a much greater area if it was to make sense, as the boundaries of development had long exceeded the County area. The second plan addressed just that.

Nonetheless, it took until 1965 for the GLC to come into existence as the first London-wide government. Even then it would only last 21 years.

This plan from the former illustrates the enthusiasm and enjoyment with which Forshaw and Abercrombie undertook the task, not only of proposing solutions to the challenges facing London after the war, but also ensuring that they understood the way Londoners worked and played.

While the pressure continued for a unified government for the wider metropolitan area between the First and Second World Wars, and strategic planning took place in the guise of the *Greater London Plan* of 1944 by Patrick Abercrombie, no action was taken until 1963, when the Greater London Council (GLC) was created.

The GLC was the result of a Royal Commission (the Herbert Commission on Greater London Government, 1957–1960) that recommended the creation of a two-tier system including a council for Greater London and 52 metropolitan boroughs as well as a separate City of London. However, by the time the London Government Act was passed in 1963, boundaries had changed again and the number of boroughs had been reduced to 34.

Between the vindictive abolition of the GLC by the Conservative government in 1986, and the creation of the Greater London Authority (GLA) in 2000, London was run by boroughs and various committees, residuary bodies and national government departments. Perhaps it would be ambitious to expect useful maps to exist to explain the complex set-up required to make such an arrangement work, but a lack of worthwhile maps followed the lack of leadership and it is only with the return of a London government in 2000 that administrative maps flourished again (in quantity if not in quality).

Perhaps encouraged by the ease of computer mapping technologies, the GLA and the Mayor's office have shown a new enthusiasm for producing and using maps to investigate and forecast the state of London. Relatively crude computer generated maps are available to show a variety of boundaries including, for example, electoral constituencies. The results of elections, almost as they are announced, are now available in map form.

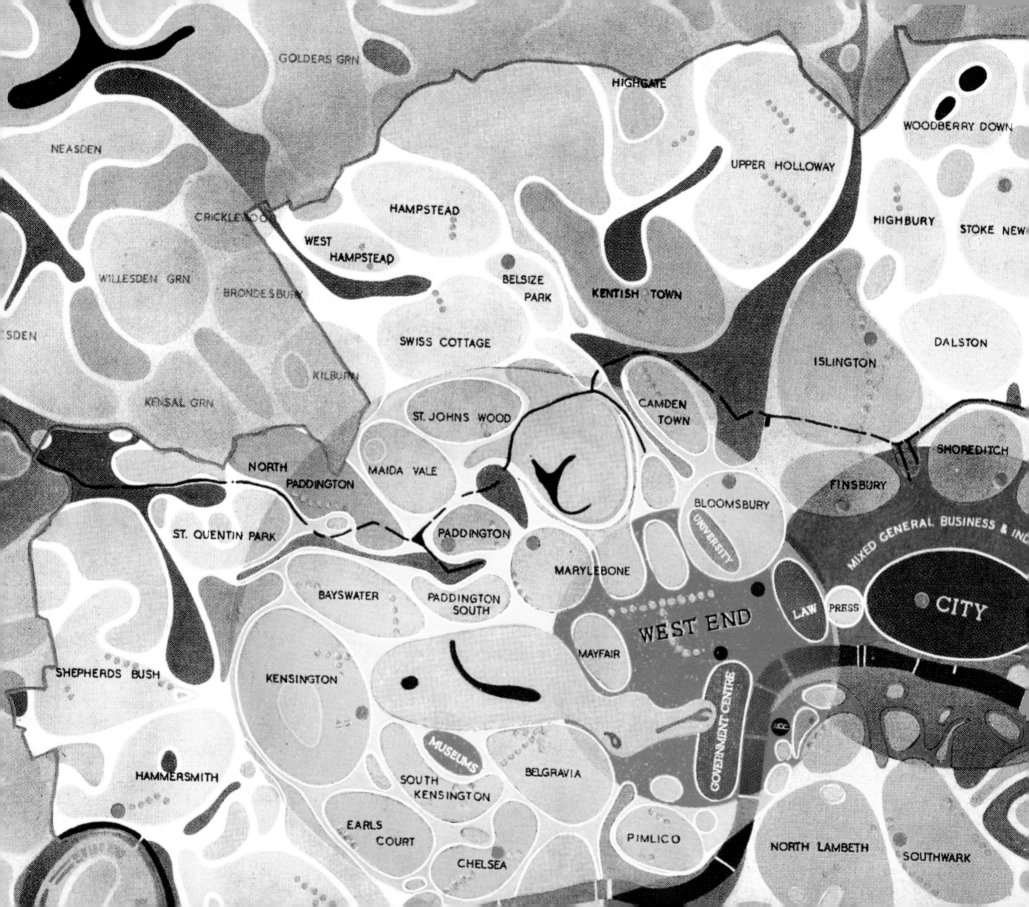

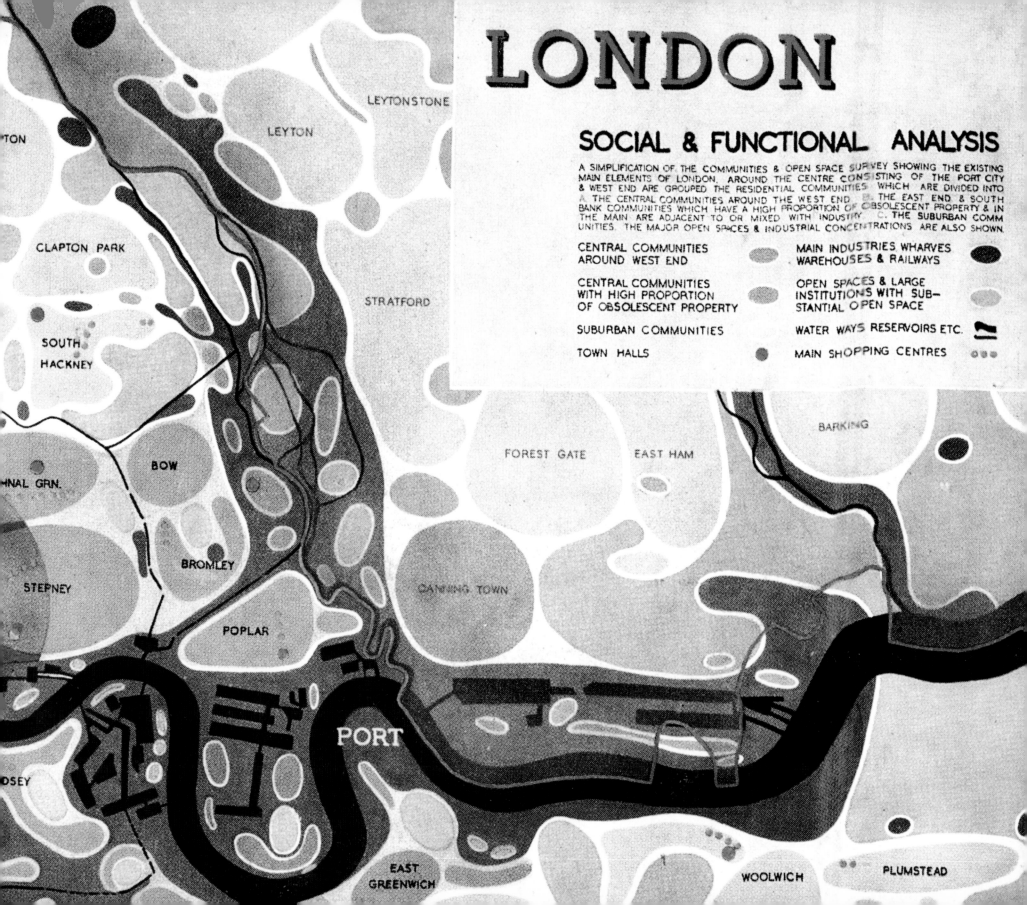

LONDON

SOCIAL & FUNCTIONAL ANALYSIS

A SIMPLIFICATION OF THE COMMUNITIES & OPEN SPACE SURVEY SHOWING THE EXISTING MAIN ELEMENTS OF LONDON. AROUND THE CENTRE CONSISTING OF THE PORT CITY & WEST END ARE GROUPED THE RESIDENTIAL COMMUNITIES WHICH ARE DIVIDED INTO A. THE CENTRAL COMMUNITIES AROUND THE WEST END B. THE EAST END & SOUTH BANK COMMUNITIES WHICH HAVE A HIGH PROPORTION OF OBSOLESCENT PROPERTY & IN THE MAIN ARE ADJACENT TO OR MIXED WITH INDUSTRY. C. THE SUBURBAN COMMUNITIES. THE MAJOR OPEN SPACES & INDUSTRIAL CONCENTRATIONS ARE ALSO SHOWN.

CENTRAL COMMUNITIES AROUND WEST END

CENTRAL COMMUNITIES WITH HIGH PROPORTION OF OBSOLESCENT PROPERTY

SUBURBAN COMMUNITIES

TOWN HALLS

MAIN INDUSTRIES WHARVES WAREHOUSES & RAILWAYS

OPEN SPACES & LARGE INSTITUTIONS WITH SUB-STANTIAL OPEN SPACE

WATER WAYS RESERVOIRS ETC.

MAIN SHOPPING CENTRES

LEYTONSTONE

LEYTON

CLAPTON PARK

STRATFORD

SOUTH HACKNEY

BOW

HNAL GRN.

FOREST GATE

EAST HAM

BARKING

STEPNEY

BROMLEY

CANNING TOWN

POPLAR

PORT

OSEY

EAST GREENWICH

WOOLWICH

PLUMSTEAD

ATTACK AND DEFENCE

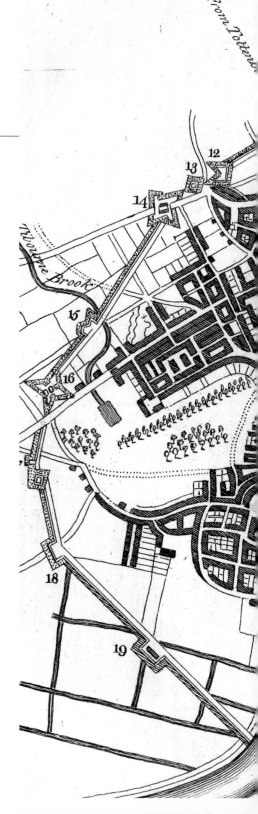

From Totten...

London has come under attack in many ways over the centuries: from insurrections and civil disturbances to enemy bombing raids. Maps have probably been made to describe each and every one of these attacks, along with an even greater number showing contingency plans for the defence of the city. The secretive nature of revolutionary, military and intelligence organisations has, unfortunately, resulted in very few of their maps surviving destruction or becoming available for public scrutiny. Those few that are viewable tend either to be immediate 'after-the-event' records of destruction and defensive action, or plans that have emerged from would-be attackers following their eventual defeat.

The governors of eighteenth century London were anxious enough about the city's defences to reproduce a 1588 survey describing the precautions taken in the Thames against the Spanish Armada, as well as a map showing the defences of the City of London in 1643 during the Civil War.[5] Neither set of defences was ever tested. But they were clearly thought significant enough to revisit and examine many years later.

Invasion maps, at least those on record, start with that most relatively gentle of regime changes—despite the fear, panic and rioting—the arrival of William of Orange in London, on 27 December

1688. The map itself is hardly military in nature and shows neither the chaos among the court of Stuart James II—as he prevaricated about fleeing before his escape to France—nor the enormous shift in British politics towards constitutional stability that this last successful invasion of the British Isles triggered.

The civil disorder best recorded in mapping is the Gordon Riots of 1780. This violent, anti-Catholic moment of mob rule raged for five days across London and paralysed the authorities until 10,000 troops were brought in to quell the destruction. Maps of the time record troop movements and encampments but show little interest in the riot itself. The lessons to be drawn from the events were clearly intended to inform future military commanders and not to help anyone hoping to instigate future insurrection.

Although riots have continued to break out regularly in London —with large numbers on the streets in support of Queen Caroline in the years 1815–1821, the Reform Bill in 1832 and the Chartists on Kennington Common in 1848, and in Hyde Park in both 1855 and 1866—mapmakers were invisible, unlike the engravers of popular illustrations who recorded the events in great detail. In 1926, however, secret contingency maps were prepared to deal with the General Strike.

City Defences (1643), 1738
George Virtue

London has rarely been attacked, but that hasn't prevented concern that it might and will be. During the Civil War, defences were erected for parliament by huge groups of ordinary Londoners, up to 20,000 on any given day, in case of attack by the royal forces. The volunteers threw up earth ramparts of five metres high and 29 kilometres long, which ran between 24 large wooden forts and completely encircled the city. In the event, they were never tested or needed. London's trained bands fought the King's army at Gloucester and then again at Newbury, preventing attack on the capital.

This is a retrospective plan prepared by George Virtue almost 100 years later, for publication in William Maitland's *The History of London*, describing the fortifications. By that time, all traces of the ramparts and forts had disappeared.

Image courtesy of David Hale/MAPCO.

Invasion route of William of Orange from Hounslow into Knightsbridge, 1688

No defences were mounted against William of Orange (William III) when he marched into London in 1688 (although unrest was feared in the city in the wake of a violent, but brief, outbreak of anti-Catholic riots earlier in the month) and his arrival was generally perceived as the chance for stability and peace. This is the second sheet of a map marking his progress from Hounslow into London, most likely prepared by (or for) one of his retinue who published them in a book shortly afterwards.

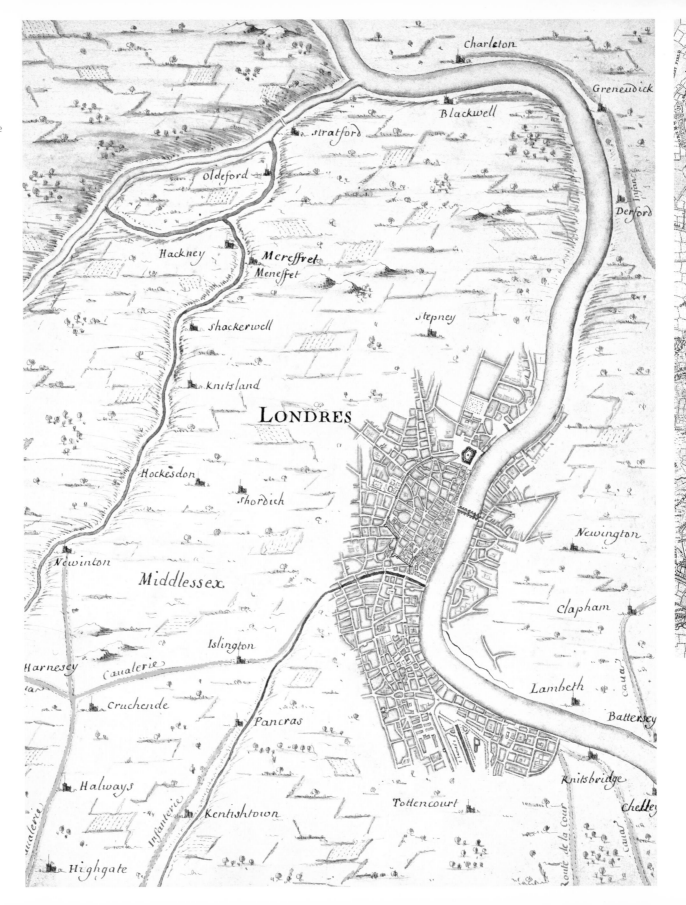

Disposition of troops during and after the Gordon Riots (1778 to 1782), 1780
John Rocque

In 1778, a time of revolutionary fervour (the American revolution had just finished and one was shortly to commence in France), a mob of 60,000—enraged by the liberalising Catholic Relief Act—gathered in St George's Fields in south London to hear Lord George Gordon speak. His emotive harangue set them off on five days of rioting, attacking parliament, foreign embassies, the Bank of England and a gin distillery as well as prisons and police offices. Eventually, the City Militia and the Honourable Artillery Company took action by employing 10,000 troops to bring the city back under control—killing 285 rioters in the process. The shock of this event—the lack of restraint of the troops as much as the actions of the rioters who had taken care not to physically hurt anyone—was profound and created a conservative backlash. The troops stayed in position for many years after. The annotation on this map by John Rocque is one of several records of the disposition of the troops. There is no equivalent map of the rioters' movements.

Bomb damage Map—No. 10, Holborn, Farringdon, St Paul's and others, 1945

Despite a few Zeppelin attacks on London in the First World War, the city's security was seriously breached for the first time during the Second World War. Sustained and heavy bombing attacks—the Blitz—were made on London between 1940 and 1941 when over 19,000 tons of explosives were dropped on the city. A period of relative calm lasted until the 'Little Blitz' in 1944, with the arrival of the first V1 pilots who carried bombs and V2 rockets. The attack on London finally ended on 27 March 1945. Throughout this period, the LCC kept meticulous records of bomb damage, which they marked in a range of colours (black: total destruction, purple: damaged beyond repair, dark red: seriously damaged, doubtful if repairable, light red: seriously damaged but repairable at cost, orange: general blast damage, yellow: minor blast damage, light blue/green: clearance areas, small circle: V2 bomb, large Circle: V1 bomb).

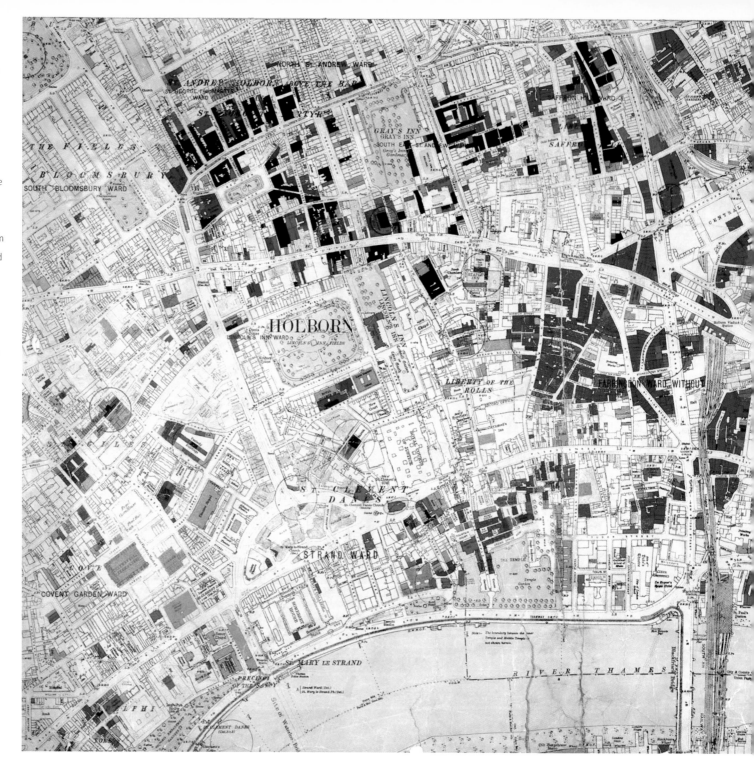

Possibly, the intention was to obliterate all trace of these maps, but surviving correspondence reveals that existing military maps were adapted for use during 1926, and that they "proved invaluable, not only to the war office, but to other government departments interested in the organisation of London at that period".[6] One version of this collection of maps was later issued as the County of London Map in 1933, despite obvious unease from its creators, the Ordnance Survey, whose director described it as "cartographically hideous".[7]

Wars, on the other hand, took place overseas, well away from London—or at least they did so before the First World War. During the Second World War, bomb and fire damage was mapped more prolifically, with the London County Council (LCC) producing a large-scale set of war damage maps in 1945. The bombing was a traumatic occasion for London and its effects have frequently been compared to that of the Great Fire of 1666 in the city's history. The result was certainly as significant to the city's planners and theorists, as they attempted to construct out of the ashes and bombed remains, a new version of London for their own time.

With the benefit of hindsight, we also have access to the maps of those that have attacked the city. Maps exist from 1944 showing the locations of V1 missiles hits, although it is now believed that they were deceived by a double agent, 'Garbo' (Juan Pujol-Garcia), in thinking their aim was far more successful than in reality. These maps, with their targets clearly marked over everyday London, have a horrifically chilling intent about them.

The latest maps to have emerged from a potential aggressor are Soviet military maps. These cover the whole world at a variety of scales, with individual towns mapped at even greater detail. Such maps date from as late as 1985 and were compiled to a high degree of accuracy, probably from a variety of sources; Ordnance Survey maps, guidebooks, street maps and even confidential Admiralty charts cross checked with high-level aerial photographs. These maps have some of the qualities of 'what if' fiction, representing an alternative reality under Soviet rule and the imaginary existence of a potential, predominantly Russian-speaking, London.

That only a tiny number of military and espionage maps are available is entirely predictable and, as they generally represent a very particular and peculiar outsider's view of the character of London, also regrettable. They offer a different perspective of London, which are made for purposes at odds with our own uses and experience of the city.

TRANSPORT AND TRAVEL

Travel and maps have an intimate history going back many centuries, if not millennia. Some of the very earliest maps are travel itineraries including, in the case of London, the Peutinger Map, a thirteenth century copy of a Roman original. Although missing one final sheet, this map shows the main road network, the *cursus publicus*, running from the western part of the empire—including eastern Britain and London—to Rome and then on to Constantinople and Antioch, with distances given between each stop on the way. Similar in purpose are the thirteenth century pilgrimage maps of Matthew Paris, a Benedictine monk living in St Albans Abbey, which were bound into his book of history, the *Chronica majora*, circa 1250s. His maps take the reader on a journey to Rome, via the major towns and cities of Europe—each of which are represented schematically and linked by short, straight, purposeful lines.

Strip maps, with the same design premise, were published by John Ogilby in his *Folio Britannia*, 1675, depicting the post routes of England and Wales and those leading out of London's centre into the outlying suburbs and beyond. Strip maps, though highly functional, primarily show the road and its landmarks—particularly the distances between towns—rather than the wider world beyond. There is no encouragement to wander off such routes or any opportunity for serendipity. Once one has diverted from the journey to some other tempting destination, the map abandons one to fate.

Travel maps, easily the great majority of maps produced annually, are very different to those plans that record and explain what is happening on the ground. Their primary concern is with the journey that takes place between points—they rarely include interesting detours. But, despite such purposefulness, the map of record is usually the preferred choice for virtual, tabletop exploration and provides greater richness and satisfaction, as well as diversion, when compared to the map designed to facilitate travel.

The travel map's relationship with physical reality can be very tenuous or, in its purest form, barely exist at all. Harry Beck's London Underground diagram is closely related to both the Peutinger and Paris maps by their combined disregard for distance and direction—topology is necessary; topography can be temporally ignored. On the Tube map, only the abstracted line of the Thames represents anything very tangible or geographically real. A transport map can choose to make its own reality and may prefer, if it does have to make a connection to the real world, to collude with systems of roads signs or symbols in place of more robust and permanent landmarks.

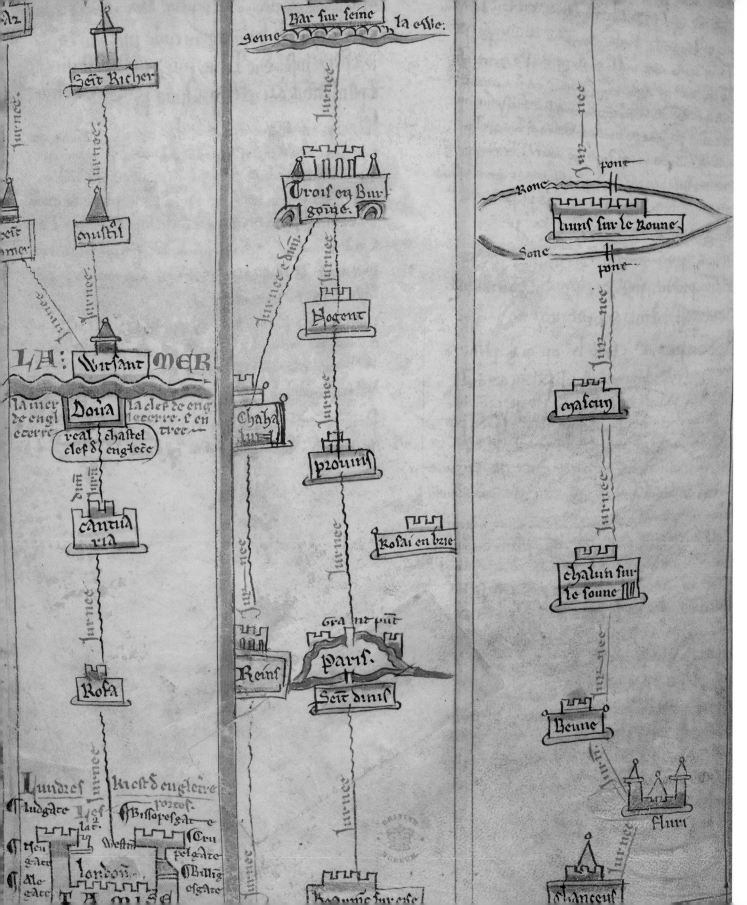

Itinerary from London to Chambery from
the *Book of Additions*, circa 1250–1254
Matthew Paris

Matthew Paris, an English monk and official chronicler at
the Abbey of St Albans, produced a vast range of materials
including histories, notebooks of miscellanea collected
information and maps, and his early representation of the
British Isles has become a familiar image. This page starts
at London Bridge (*pons Lond*) at the bottom of the page,
crosses the Thames (*Tamise*) and then onto Canterbury
and Dover before crossing the sea (*la Mer*) into France.

The style is similar to contemporary linear maps of
motorways—particularly satellite navigation devices
with their 'heads up' displays. In some cases, Paris
offers alternative routes but generally, just as with the
cursus publicus, you are not expected to deviate from
your journey.

© The British Library Board. All Rights Reserved.

Plan of London, circa 1875

John Bartholomew

By 1875, two London Underground lines were
well established, The Metropolitan Railway (the
Metropolitan Line) and the Metropolitan District
Railway (the District Line). They were privately run
and, although relations between the two companies
were sour, they did manage to share some stations and
tracks, on what is now the western end of the Circle
Line (between Gloucester Road and Edgware Road via
Paddington). This map, prepared by Bartholomew for
the publisher and newsagent, WHSmith—a company
that had prospered with the rise of the railways and
their chain of station news stands—shows the extent of
the railway system entering the city.

Image courtesy of Ashley Baynton-Williams.

Following pages:

**County Courts: Bus and Tram Routes from
Bacon's *Atlas of London*, circa 1900**

George Bacon

Trams made their first appearance on London's streets
in 1869, following approval from Parliament for three
experimental lines. By 1870, their success meant
promoters were clamouring to build and install trams
throughout London's suburbs. Within five years there
were over 60 miles of track laid for up to 350
horse drawn trams.

   This map is curious for its combination of
transport routes with the location of County Courts.
It is likely that the cartographers were simply being
economical with the number of maps of record they
were producing, which was not particularly helpful to
attendees at the courts.

Image courtesy of Thames Water.

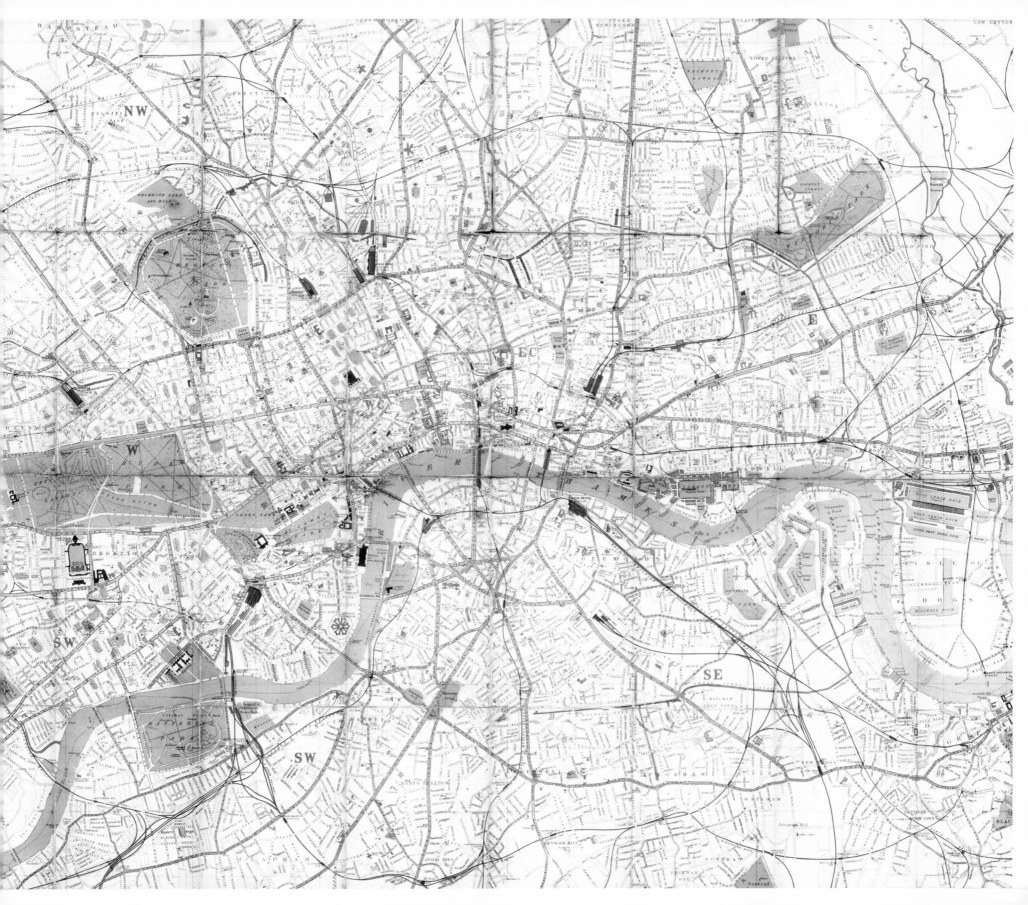

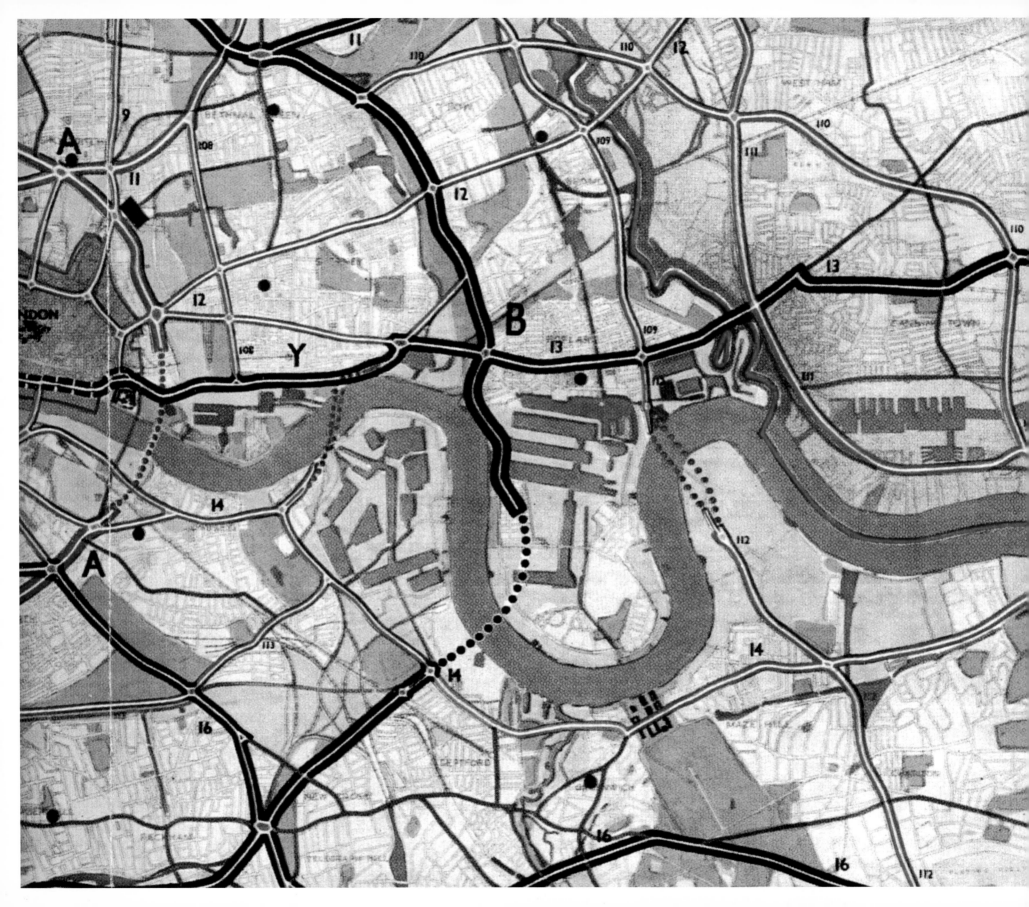

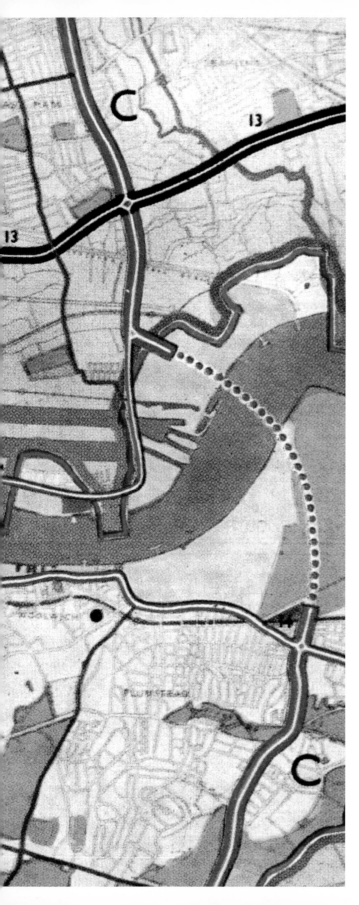

Road Plan from the *County of London Plan*, 1943
JH Forshaw and Patrick Abercrombie

By the outbreak of the Second World War in 1939, cars, motorised coaches and lorries were overwhelming a road network designed primarily for horse drawn vehicles. The need for new roads became a planning priority. Forshaw and Abercrombie proposed a series of three orbital routes, within and around London—A, B and C. Route A ran around inner London, enclosing an area similar to the 2003 Congestion Zone. The B route was a mile further out, crossing the river at Chelsea and Bermondsey, while route C follows the North and South Circular roads, already partially in place. Radial routes, including another tunnel under St James' Park and along the north river embankment linked these rings and connected the centre with the major trunk routes out of London.

The plan was logical and a brave attempt to allow London to expand. However, it would have devastated the city beyond all recognition; sweeping away vast areas of buildings to achieve a supremacy of the car. Some parts found their way into later road building plans but, on the whole, London survived.

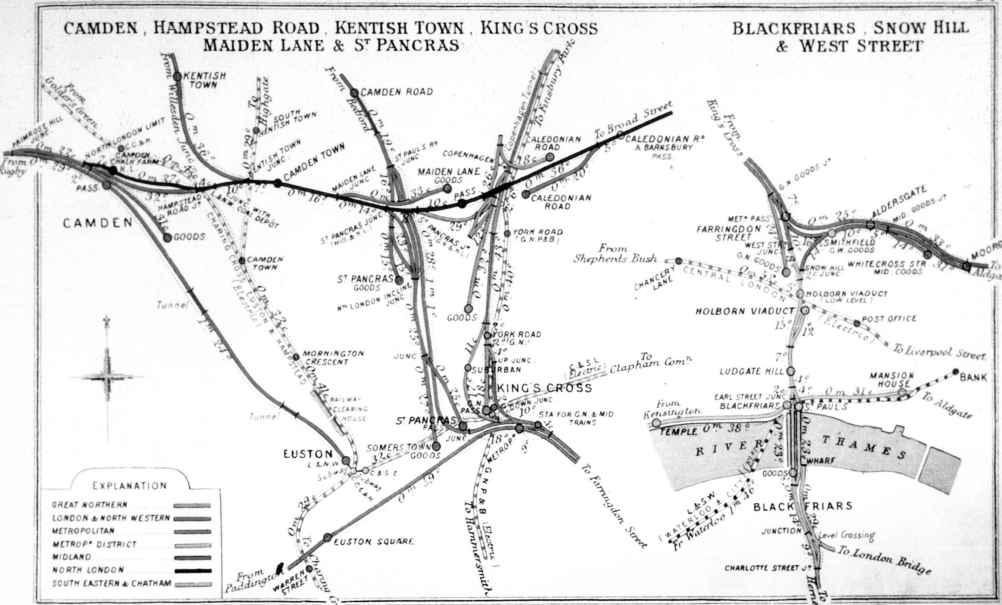

CAMDEN, HAMPSTEAD ROAD, KENTISH TOWN, KING'S CROSS
MAIDEN LANE & ST PANCRAS

BLACKFRIARS, SNOW HILL
& WEST STREET

EXPLANATION

GREAT NORTHERN
LONDON & NORTH WESTERN
METROPOLITAN
METROP? DISTRICT
MIDLAND
NORTH LONDON
SOUTH EASTERN & CHATHAM

In this instance, the journey's end becomes more important than the quality of the journey itself.

London, by its very complexity and size, demands travel maps. They make crossing the city to unfamiliar destinations possible and giving directions to other travellers relatively straightforward. Most regular users of the London Underground have a mental map of the basic system layout and need only refer to the diagram for unfamiliar journeys or unusual destinations. The diagram is also famously and almost immediately intelligible to strangers to London. It has imposed a conceptual framework on the undisciplined chaos of the streets and local centres of the city. Recent bus maps, with very different requirements, have attempted something similar, generating a hub and spoke model of London, while generic road maps tend to portray London roads as a series of rings with main, interconnecting radials. Each map system tries to deny reality in favour of an easier and more memorable, alternative structure.

The result is the difficulty many experience in navigating directly from area to area. The zone around each London Underground station becomes connected to adjacent zones by the transport system rather than by the simpler walk. Travellers will still use the Tube to move between stations that are geographically close together, despite being told that they are only a short walk apart. Drivers, on the other hand, are actively encouraged to take the orbital routes around London (M25, North and South Circulars, Congestion Zone circumference) rather than more direct routes. Transport maps, rather than broadening an understanding of London, hide and deny its complexity in favour of simpler patterns, more strident diagrams and the idea of citywide homogeneity and coherence. Their success, and ability to create their own sense of logic, have left many parts of London strangers to one another and ignore the real patterns of urban growth and topography.

Nonetheless, mapping of many forms of transport have failed to find such simplifying diagrams, often despite considerable effort on the part of mapmakers. These include maps for cycles, taxis and pedestrians. Cycle routes in London have been mapped since a craze for the activity took hold in the 1880s and 90s. The earliest cycle maps, by Letts, George Philip, Bartholomew, Edward Grove, et al, were aimed at those who used bicycles to escape the inner city during their leisure time. The concept of the cyclist-commuter comes later, and is accompanied by few specially prepared maps until the arrival of cycle routes in the 1970s and the publication of the first *On Your Bike* guides by the Kensington and Chelsea Friends of the Earth group under the editorship of Owen Watson in 1976.

Left:

Map of the London Underground system, 1908

One of the earliest combined London Underground maps, on which the various lines—all owned by different companies at this point—are shown together. It may well have been an unauthorised plan, as the Waterloo and City Railway Company had not yet agreed to participate in a joint map. The text in the upper left corner is suspiciously similar to the real logo, first seen in 1908.

The map makes a valiant attempt at creating a strong graphic identity for the London Underground map, which although approximately geographical, has been stretched east–west to distort the form of the lines. The bright colours on the black background, although striking, makes the individual station names hard to read.

Image courtesy of Mary Evans Picture Library.

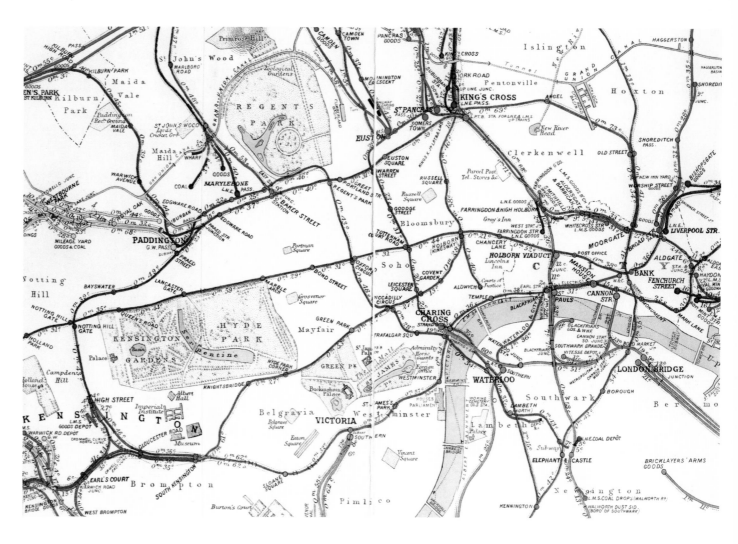

Railway Clearing House Map of Central London, detail, 1935

This map of the railway lines that were under the jurisdiction of the Railway Clearing House, gives an accurate geographical representation of this part of the London Underground network. It also demonstrates how far many of the geographical maps of this system had begun to deviate, in an attempt to make station names legible.

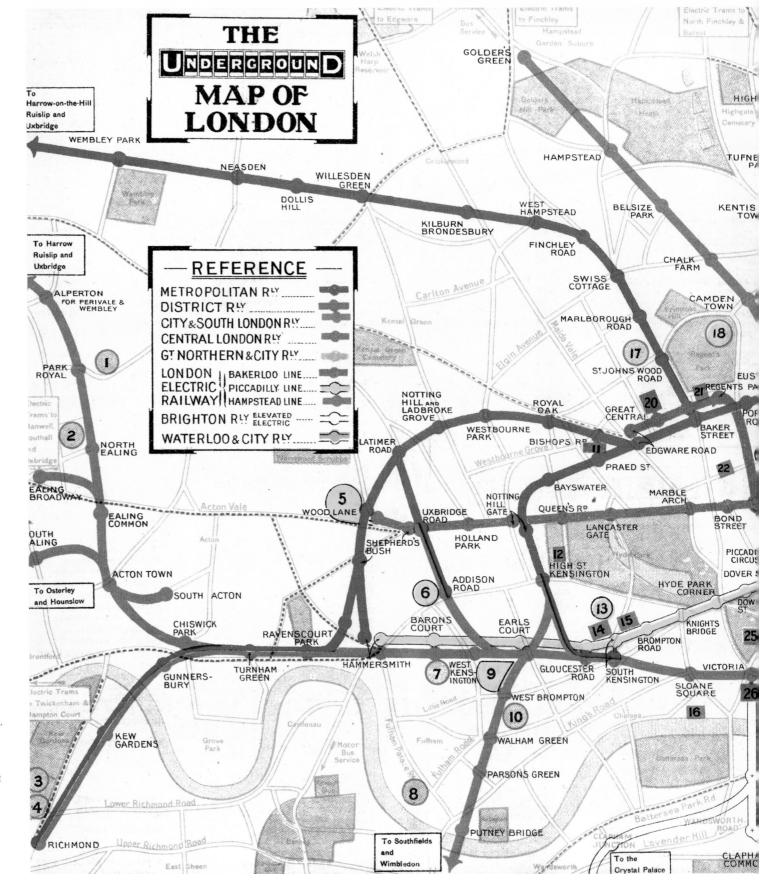

The Underground Map of London, 1911
Johnson Riddle & Co.

In the early stages of developing the London Underground map, graphic designers played with straightening out the trajectories of the various lines, invariably making it easier to depict station names. This 1911 map is more diagrammatic and vivid than most, but it cannot have found favour, as subsequent maps return to a stricter geographical adherence.

Image courtesy of David Hale/MAPCO.

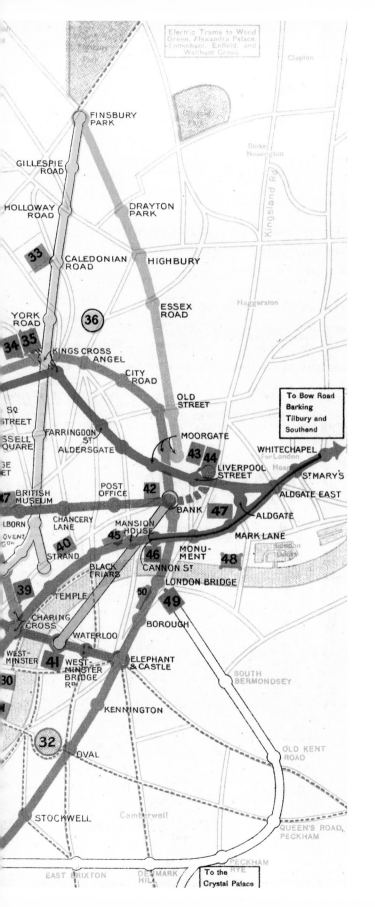

London Town, 1948–1965

In an attempt to promote rail travel into London, British Rail published this map poster in 1948, showing the attractions of the capital. For all its popularising zeal, the poster gives London a murky brown quality at odds with its intent. The image of pageantry and Medievalism is one that the city has spent much of the last half century trying to shake off.

Image courtesy of Science & Society Picture Library.

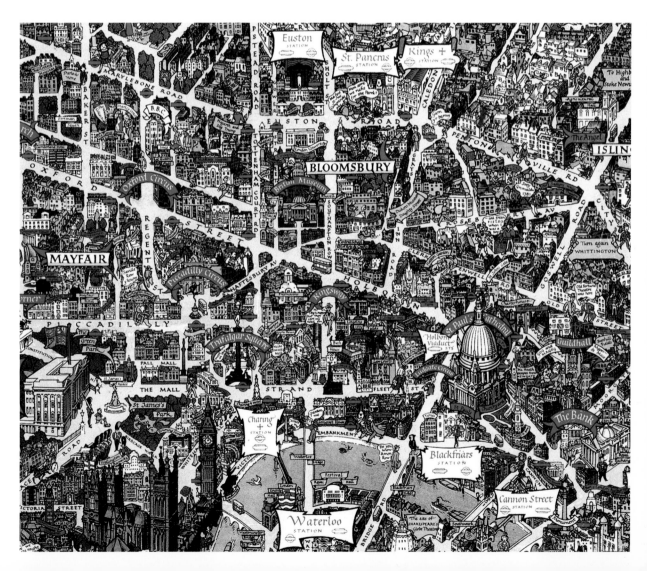

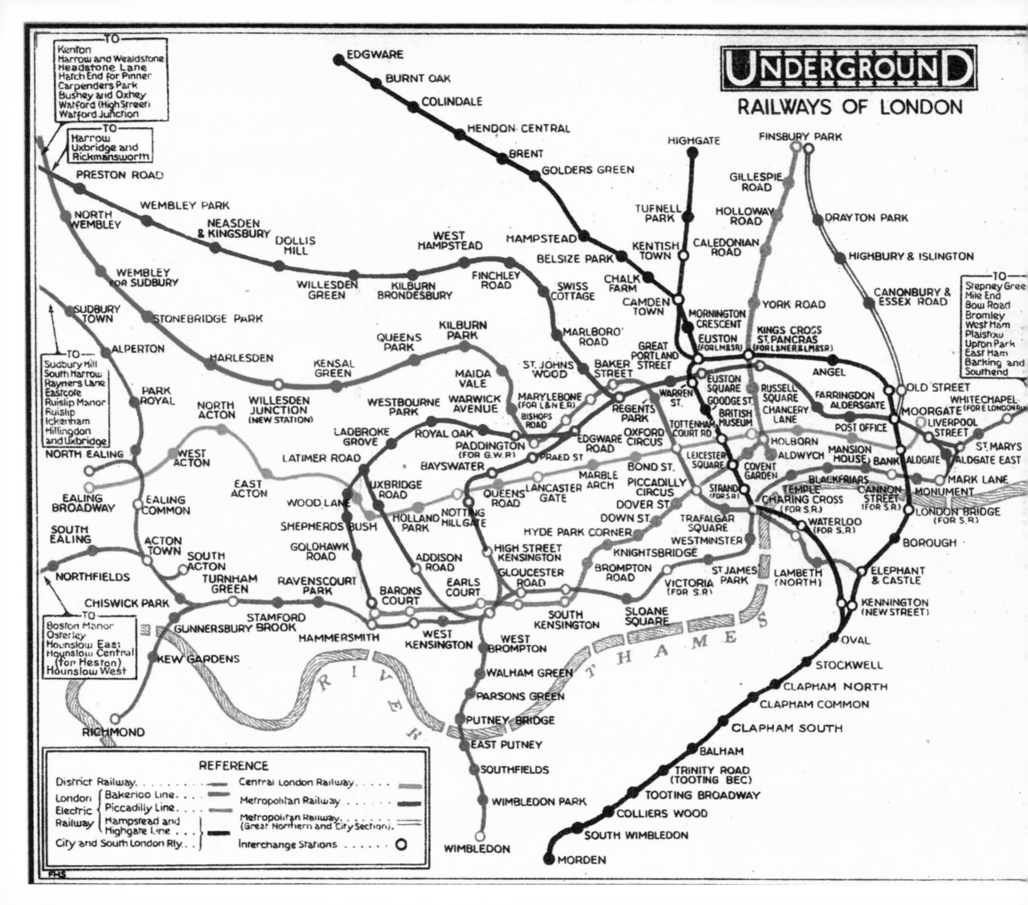

Underground Railways of London, 1927
Fred Stingemore

The second of the 11 maps drawn by Stingemore for the London Underground. In his first, he omitted all surface detail but in this version, the river reappears (remaining in place on all subsequent London Underground maps and diagrams). The layout is almost geographically correct, while the colour of each line begins to approach what we are familiar with today.

Image courtesy of Transport for London.

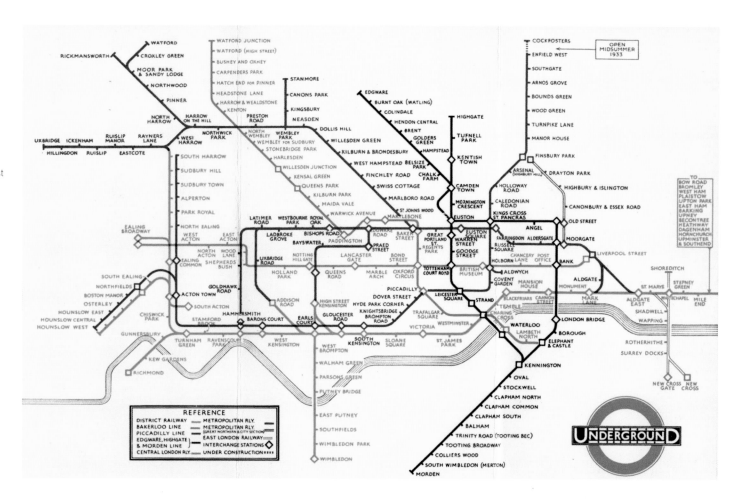

Underground Diagram, first edition, 1933
Henry Beck

The first printed edition of the London Underground diagram was on a folded card distributed among the travelling public. It had been significantly developed from Beck's original mock-up. The Piccadilly Line is the same tone of blue colour as today and interchange stations are signified by diamond markers. Most significantly, the short 'ticks' indicate the other, through stations, with the 'T' marking the terminus. The diagram was already being refined for clarity and elegance—a process that would continue over many years.

The map was enthusiastically received by the public from the outset (more so than by London Underground management) and Beck was commissioned to prepare further editions including large-scale 'quad royal'

size posters for display on the station platforms. Beck devoted himself to developing and refining his plan, as well as warding off unwanted suggestions for 'improvements' from outsiders—including London Transport managers. In 1960, London Transport commissioned a new diagram from Harold Hutchinson, without informing Beck. Hutchinson was then replaced by Paul Garbutt in 1963. The current designers of the diagram, recognising the importance of Beck's contribution, have restored much of his thinking in the system's ongoing development.

Image courtesy of Transport for London.

Modern London and its Environs, 1849
Edward Mogg

Travelling in London in the mid-nineteenth century wasn't easy. Most journeys were still done on foot or in Hackney coaches (four wheeled), hansom cabs (two wheeled) or on the new omnibus service. Omnibuses—horse drawn coaches that could carry approximately 20 people on a single deck—were introduced by George Shillibeer in 1929. They soon spread all over London, connecting the railway termini and reaching deep into the suburbs. By 1850, there were up to 1,300 such vehicles on the roads of London—a city already suffering from considerable traffic congestion.

Mogg's map, as its subtitle "The coloured lines exhibit the Omnibus routes" suggests, was designed to help travellers and commuters find their way on the expanding system, whose rapid growth must have been a cause of both new opportunities and confusion. The large, overprinted circle represents a distance of two miles from the central General Post Office.

Image courtesy of the Guildhall Library, City of London.

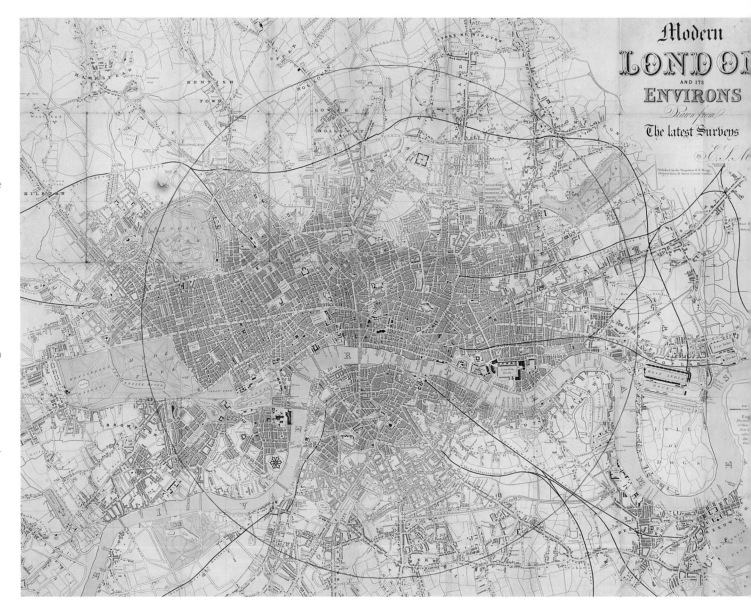

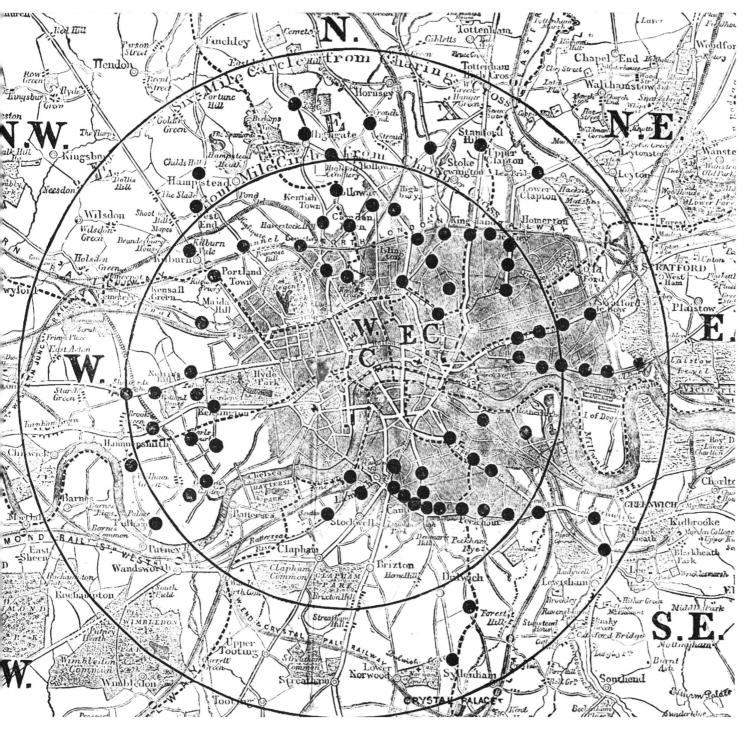

London Toll Gates from
The Illustrated London News, 1857
John Dower

In order to deal with maintenance costs, roads and some of the bridges around London were transformed into toll roads or turnpikes. Maps were published showing locations and rates, as well as offering useful advice as to how to circumvent the highest charges.

With the railway system offering alternative means of avoiding tolls, and a public clamour against the delay and expense, it was necessary to form a Toll Reform Committee, which was undertaken by Herbert Ingram in the mid-1850s. This map depicts London's toll roads at the time of the Committee's report in 1857. It subsequently concluded to abolish "all toll gates and bars within a radius of six miles from Charing Cross, particularly on those roads north of the Thames which are now out of debt". Within a year, Parliament acted to remove the tolls and such a system would only return to London in 2003 in the guise of the Congestion Zone scheme.

Cycling Road Maps of London, circa 1900
Edward Grove

Cycling became very fashionable in late Victorian and
Edwardian London with large numbers gathering to
parade in parks, much as previous generations had
done so on horseback. Long distance excursions were
also popular and many took the opportunity to cycle
out of London on weekends. Maps, such as this one by
Grove, helped cyclists find their way out of the centre
of the city and into the countryside.

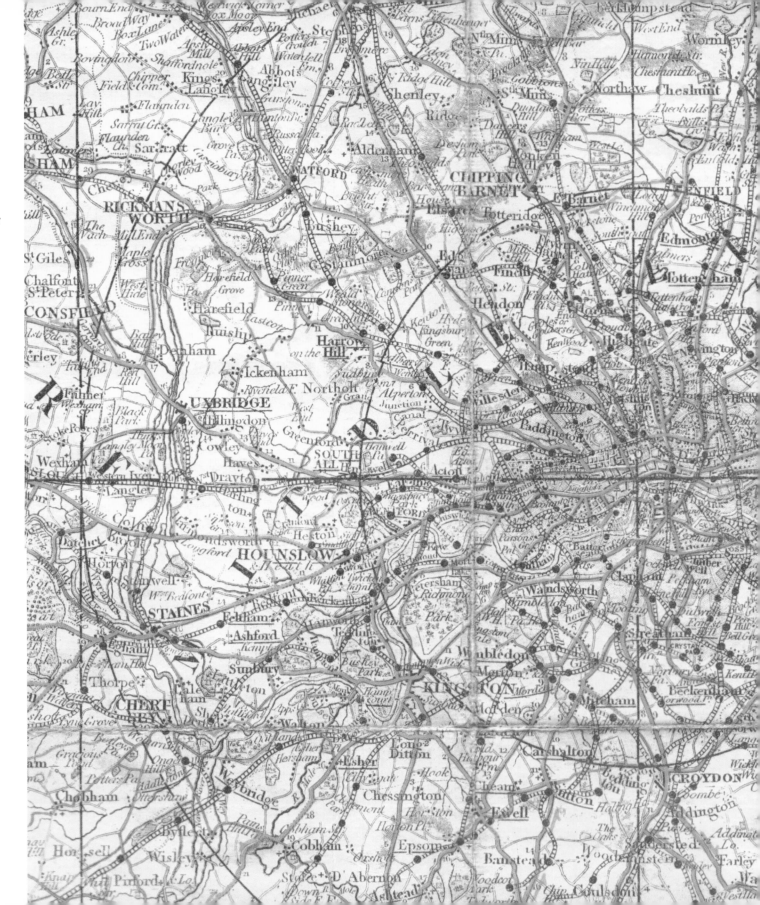

On Your Bike front cover, 1982
Owen Watson

The activity of cycling received renewed interest in the environmentally conscious 1970s. Friends of the Earth and the London Cycling Campaign would collaborate to produce a series of map guides helping cyclists to navigate through the quieter back streets of London and avoid major traffic hotspots. These maps highlighted the nascent cycle routes beginning to be provided by some of the boroughs. The guides have their forerunners in the Victorian maps of Letts and Grove and, in turn, influenced the London cycle guides issued by Transport for London which now cover over 4,000 miles of routes.

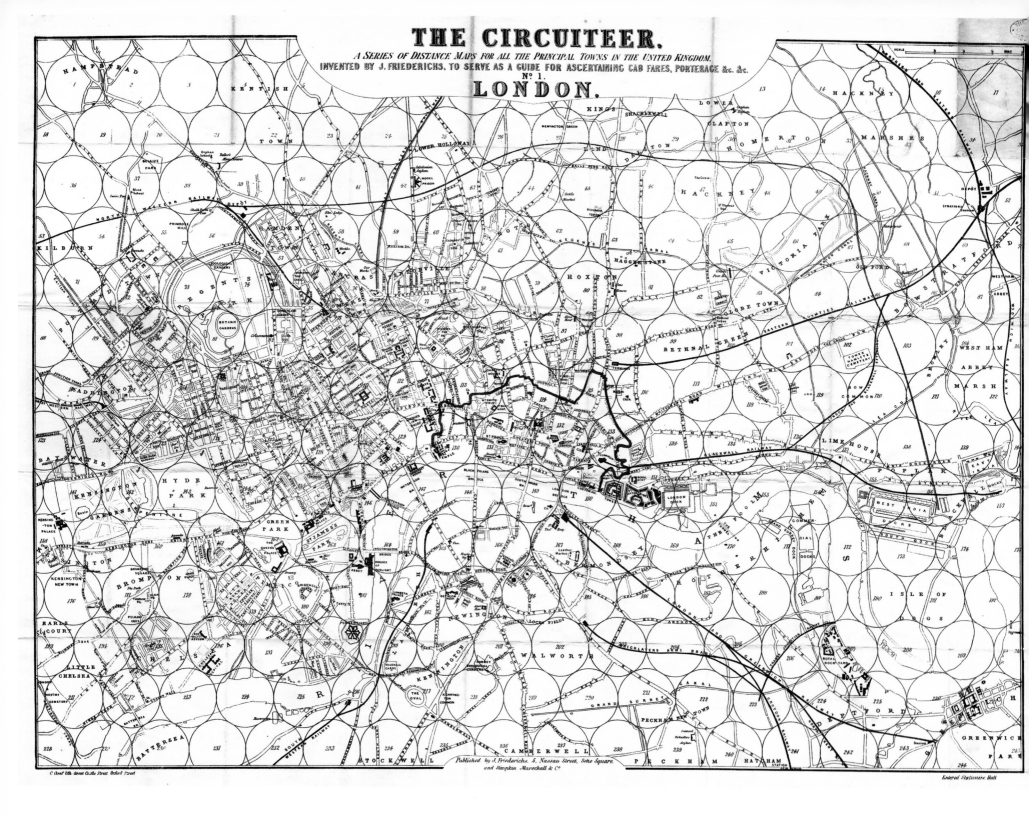

THE CIRCUITEER.

A SERIES OF DISTANCE MAPS FOR ALL THE PRINCIPAL TOWNS IN THE UNITED KINGDOM,
INVENTED BY J. FRIEDERICHS, TO SERVE AS A GUIDE FOR ASCERTAINING CAB FARES, PORTERAGE &c. &c.

Nº 1.

LONDON.

Published by J. Friederichs, 5, Nassau Street, Soho Square.
and Simpkin Marschall & Cº

The Circuiteer, 1847

(1862 edition)

Joachim Frederichs

By the mid-nineteenth century, Hackney coaches and hansom cabs were ubiquitous on the streets of London. In 1845 there were 2,500, and by 1863 this had risen to 6,800. It would rise again to 11,000 by 1988. But the number of competing cab drivers could not persuade passengers that they weren't being taken for a ride in both senses of the expression. Prices were fixed at 18 pence a mile and mapmakers were keen to help users calculate the distance of their journeys to ensure they were being charged a fair price. Maps were produced with tape measures bound into the jacket, overprinted with triangles, hexagons and, on this one by Joachim Frederichs, circles, to make counting the miles possible.

Image courtesy of David Hale/MAPCO.

More recently, cycle maps have come under the aegis of Transport for London (TfL) and the Mayor's office but they have yet to find a distinctive quality that makes them memorable rather than simply useful.

Greater efforts were made with taxi maps in the nineteenth century (before the arrival of the trustworthy taxi metre of course) in order to establish fair methods for reckoning fares. It was widely believed that taxi drivers, in Hackney in particular, regularly fleeced their customers, and so mapmakers used all their ingenuity to find ways of assisting passengers in calculating what was a reasonable charge. The methods employed usually involved elaborate griding of London maps (circles by Friederichs in 1837, triangles by Simpkin & Marshall in 1851 and squares by Mogg in 1876), which could then be used to add up distance, and as a result, fares.

Walking, in contrast, continues to fascinate the creators of maps—possibly in response to the perceived difficulty of walking through London's streets easily—and there are many new maps; monitoring, measuring and encouraging pedestrians as they find their way through the city. Both Space Syntax and Intelligent Space Partnership have developed pedestrian movement maps and there are a range of navigational schemes and map-based guides ready to assist those on foot to make sense of the chaotic signage of London and find their way from area to area. While most other transport maps are computerised and become interactive, traditional, paper-based walking maps are enjoying a renaissance at the start of the twenty-first century.

But despite other methods of visualising travel in London, the success of the London Underground diagram means that it continues to rule supreme and is endlessly adjusted and adapted to show more information, including; regional railways (TfL and British Rail, 1973), additional lines (TfL, 2006), traffic density (del.icio.us, an interactive isochronic morphing of the London Underground map to show journey times from different stations by Tom Carden, a work in progress). The Tube map now effortlessly represents London with all of its distortions of place and distance; and by doing so has been instrumental in binding together some very distant and far-flung places into the idea of belonging to the capital.

London Pedestrian Mapping, 2006
Space Syntax

Londoners have always walked, despite the many
difficulties thrown in their way by unsympathetic
road systems, and recent years have seen a number
of studies aiming to understand and facilitate walking
in the capital. Space Syntax have used the Spatial
Accessibility Model to assess and forecast pedestrian
movement patterns through London—as shown on their
pedestrian mapping diagram—based on much research
and through synthesising thousands of individual
journeys. This has also led to the development of a
map, not dissimilar to other London transport maps, to
support and encourage walking by providing a simple
memorable picture of key walking routes.

© Space Syntax Ltd.

Highly accessible areas for pedestrian movement.
These are the primary circulation routes.

Key wayfinding routes. These are important
in terms of wayfinding within the street network.

Routes that are close to, but not on, primary
routes. These are the secondary routes.

These are secluded routes. These are local routes
which are secluded from view and difficult to navigate.

Intelligent Space Map
Pedestrian Flow Model, 2006
Intelligent Space Partnership

Intelligent Space is a consultancy that provides
expertise on pedestrian movement and space use.
This map depicts a spatial analysis of central London's
pedestrian movement network, measuring the levels of
"street network accessibility" for pedestrians. The most
accessible areas for walking are shown in red, through
a spectral range to the least accessible in blue.

Image courtesy Intelligent Space Partnership Ltd.

Curved Map, 2005

Maxwell Roberts

The strict linear rules of Beck's original London Underground diagram were begging to be challenged in subsequent renditions. This example, by Maxwell Roberts, does just that and despite such reinterpretation, remains recognisable as part of the Beck/London Underground family.

Image courtesy Maxwell Roberts.

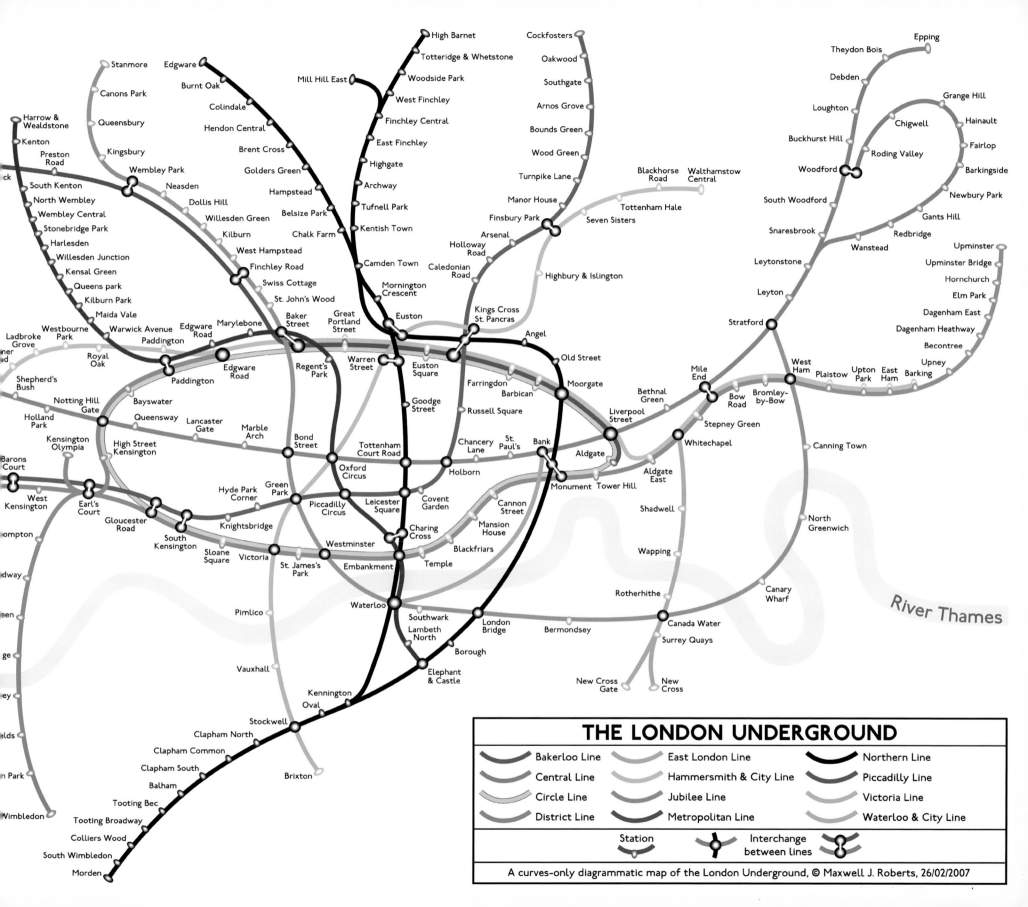

THE LONDON UNDERGROUND

Bakerloo Line		East London Line		Northern Line	
Central Line		Hammersmith & City Line		Piccadilly Line	
Circle Line		Jubilee Line		Victoria Line	
District Line		Metropolitan Line		Waterloo & City Line	

Station   Interchange between lines

A curves-only diagrammatic map of the London Underground, © Maxwell J. Roberts, 26/02/2007

Health, Water and Waste

The street, and any convenient ditch, used to be the best place to get rid of London's unwanted waste and detritus. From there, it could flow or be washed into a stream or river, and eventually out into the Thames. Understandably this was illegal. Richard II decreed in the fourteenth century that "none shall cast any garbage or dung or filth into ditches, waters or other places". Nonetheless John Stow, writing in 1598, laments that despite continual efforts to keep the Fleet river running and clean, and "much money being therein spent, the effect failed, so that the brook, by means of continual encroachment upon the banks getting over the water and casting of soilage into the stream, is now become worse cloyed and choked than ever it was before".[8]

Household waste was intended to be collected in cesspools in the basements of houses and thirteenth century regulations specified the size and construction of these 'chambers'. The contents were then collected by 'nightsoil' men for use as manure and, from the sixteenth century on, for extracting saltpetre for making gunpowder. Such cesspools were emptied by cart by 'rakers' who would sell it on to local farmers, resulting, in turn, in stinking piles of excrement along the approach roads into London. Sewers existed, which drained into rivers and streams, but they were intended for surface water only and local acts—following on from the 1531 Bill of Sewers—made it illegal to discharge any household waste into them. It was not a system that was pleasant, effective or healthy. As Pepys reports in his diary on 20 October 1660:

> This morning one came to me to advise with me where to make me a window into my cellar in lieu of one that Sir W Batten had stopped up; and going down my cellar to look, I put my foot into a great heap of turds, by which I find that Mr Turner's house of office is full and comes into my cellar, which doth trouble me; but I will have it helped.[9]

This was a system that was accepted for centuries; until an ever increasing population, combined with the widespread use of the flushing toilet (invented by Joseph Bramah and patented in 1778)—which drained straight into the storm water sewage system and out into the rivers—brought the system to a point of total collapse.

The crisis came to a head, at least as far as Parliamentarians were concerned, in June of 1858 when the stench of raw sewage rising from the banks of the Thames, through the windows of the relatively new Palace of Westminster (completed by Barry and Pugin in 1852) became intolerable. The reaction to the Great Stink was immediate and a bill was introduced on 15 July to extend the powers of the Metropolitan Board of Works (MBW) to act on the issue and, vitally, to raise 33 million

MICROCOSM dedicated to the London Water Companies { BROUGHT FORTH ALL MONSTROUS, ALL PRODIGIOUS THINGS. HYDRAS, AND GORGONS, AND CHIMERAS DIRE. Vide Milton

MONSTER SOUP commonly called THAMES WATER, being a correct representation of that precious stuff doled out to us !!!

Monster Soup, 1828
William Heath

Thames water was foul and dangerous. The sewers of an ever expanding London—originally only intended as rainwater drains—were depositing increasing amounts of bodily waste directly into the river (often well above the out-takes of water supplies back to the houses of the capital). The problem, in a de-regulated city with no central authority, was getting anyone to do anything about the problem.

William Heath's cartoon of 1828, 30 years before the Great Stink occurred, is only one of many scathing attacks on the death-dealing nature of the Thames. The view of the bug-infested water in the microscope could be a map in itself—depicting a disease-ridden city, content to endlessly re-cycle and re-ingest its own waste.

Image courtesy of Thames Water.

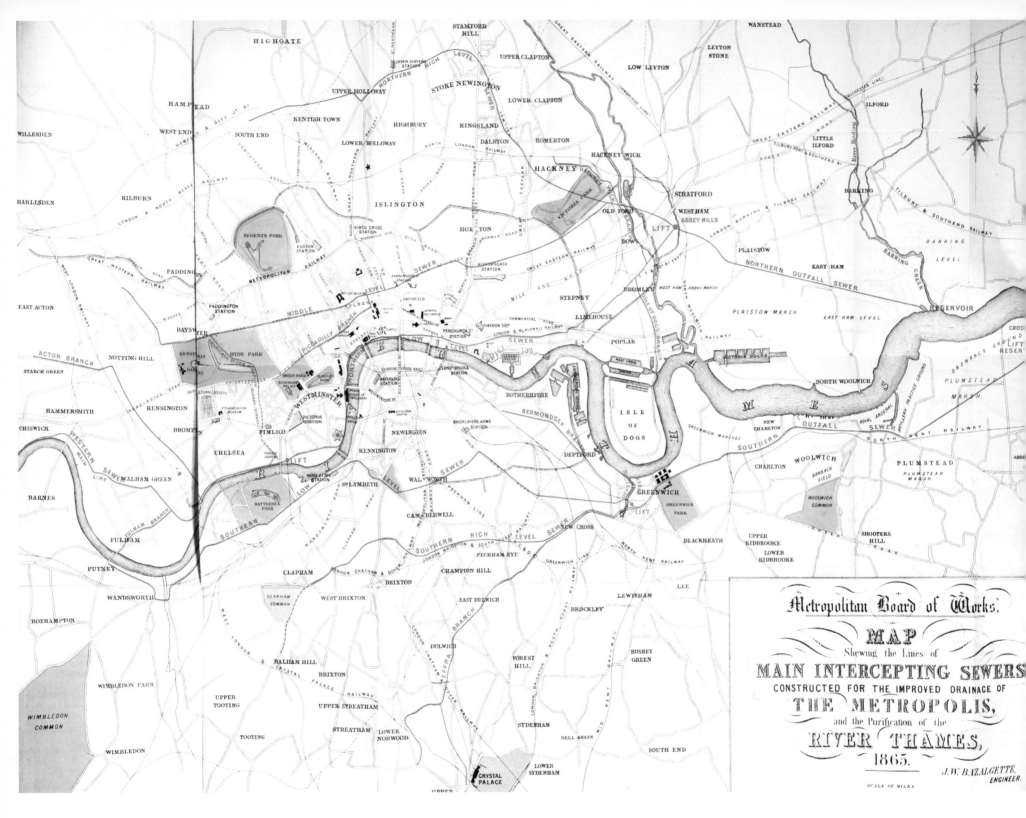

**The Lines of Main Intercepting Sewers,
constructed for the Improved Drainage
of the Metropolis, 1865
Joseph Bazalgette**

The creation of the MBW gave its chief engineer, Joseph Bazalgette, the authority and resources to design and construct his great system of sewers—a system that, with modifications and improvements, continues to work today. Bazalgette's plan, as shown in this drawing, is for a series of sewers running either side of the Thames. The sewers were aided, where necessary, by pumps and lead to storage tanks on the tidal estuary at Beckton, on the north bank, and Crossness on the south. The works also involved building the great Victoria and Albert Embankments along the river, partly to contain the low-level riverside sewer but also to encourage the water to flow more effectively.

© Guildhall Library, City of London/
The Bridgeman Art Library.

pounds to do so. The act became law on 2 August. It may be that the MBW had allowed the problem to build up over that hot, dry summer so that they could mobilise political will and authority behind the plans for a sewerage system that they and many others had been planning for years. If so, they succeeded. The passing of the act meant that the MBW could get work started on the great system of outfall sewers, as already designed by its chief engineer Joseph Bazalgette, in early 1859.

Bazalgette's plan, in part based on 137 proposals invited from and put forward by the public in 1849 (and also a very similar proposition submitted by the artist John 'Mad' Martin in 1831) was for a system of intercepting sewers, north and south of the river that led to pumping stations at Abbey Mills and Crossness respectively, which would store and then discharge the effluent into the river during periods when the tide was flowing out to sea. In addition, embankments would be built along the banks of the river to encourage it to flow effectively. It was a scheme that would take 20 years to build and that would, above anything else ever attempted, transform the map of London and make possible its continued enlargement. It is a work of engineering that, despite substantial renewal works, is still effectively in use today.

The pre-1850 drainage system and its incipient collapse had another far more dangerous effect than an intolerable smell from the river. The river was the source of drinking water for large areas of the capital. Basement cesspits were overflowing, infecting other water supplies and in the mid-nineteenth century, London suffered from four major cholera outbreaks (from 1831–1832, 1848–1849, 1853–1854 and in 1866) that killed over 37,000 people. The cause of cholera was unknown but it was widely believed by scientists, media and public alike to be bad air or miasmas—in other words: 'the smell'. Some campaigners, including the very forceful Edwin Chadwick, strongly recommended improving the drainage in order to remove the miasmas and hence the cause of infection. Bazalgette's drainage scheme was implemented on this assumption.

However, Dr John Snow, another campaigner, believed otherwise. In 1849, he published the paper "On the Mode of Communication of cholera", which argued that the disease was spread in drinking water infected by sewage. He also suggested that the practice of draining human waste into the river to be later drunk by thousands of others had significantly exacerbated the epidemic of 1848–1849. But his was a view that contradicted the scientific orthodoxy and was largely ignored.

**Cholera cases around the Broad Street Pump
from "On the Mode of Communication
of cholera", 1855
Dr John Snow**

Snow, using the weekly mortality figures compiled by
William Farr, London's Registrar General, recorded
every incidence of cholera that occurred in London
in 1855. This map wasn't the only one to depict such
data, but it was the first that helped identify cholera as
a waterborne disease. Snow experienced widespread
derision at such findings, including an editorial in the
Lancet, when described as a hobbyist who had "fallen
through a gully-hole and has never since been able to
get out again". By the time of the next serious outbreak,
opinions were revised—the same editorial declaring him
the benefactor whose work "has enabled us to meet and
combat the disease". This map was Snow's means of
proving his case to the disbelieving world.

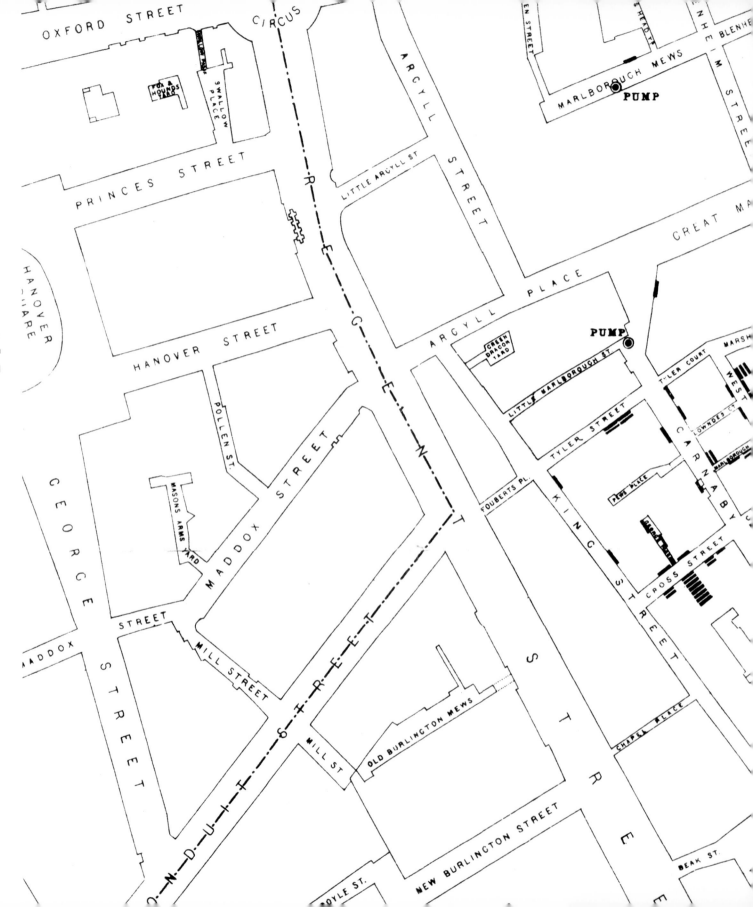

METROPOLITAN BOARD OF WORKS

SHOWING THE SEVERAL DISTRICTS OR PARISHES AND THE
PORTIONS OF EACH DRAINING INTO THE HIGH LEVEL, MIDDLE LEVEL
AND LOW LEVEL SEWERS RESPECTIVELY ALSO THE TIME
OCCUPIED BY THE SEWAGE IN TRAVERSING THE SEVERAL SEWERS.

Note *The area draining into the High level sewers is colored red, that draining into the Middle level sewer on the North, and into the Outfall sewer on the South is colored green, and that draining into the Low level sewers is colored blue.*

The thin numbers (thus 1) refer to the several Districts or Parishes. The heavy numbers (thus 2) along the sewers, show the time in hours occupied by the sewage in passing from that point to the outfall reservoirs at Barking

The districts surrounded by hatching thus ___ are those in which the Population is decreasing.

HIGH LEVEL SEWER

HACKNEY WICK BRANCH

ABBEY MILLS PUMPING STATION

OUTFALL SEWER

MIDDLE LEVEL SEWER

PICCADILLY BRANCH

ISLE OF DOGS BRANCH

BARKING

CROSSNESS

LOW LEVEL SEWER

BERMONDSEY BRANCH

OUTFALL SEWER

Hammersmith Bridge

LOW LEVEL SEWER

HIGH LEVEL BRANCH

EFFRA BRANCH

Scale 1 inch equal 1 Mile.

Plan of the District of the Metropolitan Board of Works, circa 1886

London's water had long been drawn from the Thames. Water wheels installed in the arches of London Bridge pumped supplies through elm trunk pipes to the City (from 1581) and Southwark (from 1761). A mechanical pump, initially horse and later steam driven, near the site of the present Charing Cross Station supplied (from 1691) areas of Westminster. The related wooden pump tower, the York Buildings Company, is very prominent in Canaletto's 1750–1751 view of the Thames from the terrace of Somerset House. North of the Thames, London was supplied with water by the New River Company, which drew its water supply from the Amwell, 40 miles along the newly created 'New River'. Further west, the Chelsea Water Works Company also supplied water to Westminster (from 1723 onward) via reservoirs in both Green Park and Hyde Park. There were numerous other smaller companies supplying outlying areas of London. At the beginning of the nineteenth century, these companies were engaged in a pitched and ruthless battle for customers, resulting in the General Agreement; to carve London up into company monopolies. But, given the sources of supply and the leaking pipes that were only filled during the few hours that water was made available (approximately for two hours, three days each week), the water was still unsanitary and life-threatening—the typical lifespan of a Londoner remained short.

The state of the water supply at the beginning of the nineteenth century was generally agreed to be scandalous. A pamphlet in 1827 pointed out that, in many cases, water inlets were located directly opposite sewage outlets. A Royal Commission in 1928 recommended that the intakes lower down the Thames should be replaced by sources further upstream. The findings were ignored but companies did start to filter their supplies and remove the larger impurities. It was only the agitation of Edwin Chadwick in his "Report on the Sanitary Condition of the Labouring Population", 1842, and the impact of the 1848–1849 cholera outbreak, that forced changes and eventually led to the 1852 Metropolitan Water Act. This act required companies to implement radical changes to their supplies which resulted in far cleaner water and, in turn, to a healthier population.

But before change was comprehensively implemented, either to the drains or the supply of water, cholera broke out twice more. During the 1853–1854 outbreak Snow, no doubt on the look out for evidence to prove his earlier theory, collected compelling findings based on the evidence of cholera surrounding a well in Broad Street, Soho. In presenting these findings, he transformed cartography. Snow prepared a map of the area around the Broad Street pump, marking individual cases of cholera and, as a result, showed that those who used the pump had a far higher chance of contracting the disease than the residents

London Water supply Ring
Main Schematic, 2007

London water, much of which is still extracted from the Thames and the River Lea, was supplied by cast iron pipes from 1746 onwards. With the continual corrosion of this system, Thames Water constructed a Ring Main at a depth of approximately 40 metres below ground level from 1988 to 1993. Water enters the ring from the water treatment works and can flow in either direction to maintain pressure and supply at any point within its looping plan.

The Ring Main is approximately 51 miles in length, with some 12 pump-out shafts supplying water across London. The system supplies in the region of 1,300 megalitres of water a day.

Image courtesy of Thames Water.

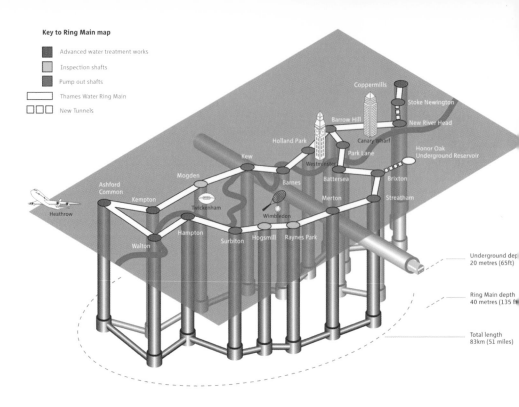

Key to Ring Main map

- Advanced water treatment works
- Inspection shafts
- Pump out shafts
- Thames Water Ring Main
- New Tunnels

Underground depth 20 metres (65ft)

Ring Main depth 40 metres (135 ft)

Total length 83km (51 miles)

of surrounding streets with alternative supplies. The pump handle was removed and investigations showed a closely passing sewer was leaking into the well. Overnight, the map became a means of concentrating information and revealing patterns that were otherwise invisible. It also demonstrated how maps could become a persuasive, campaigning medium. Yet, despite his work, few were prepared to believe him or change their faith in the conventional miasmic theory.

In a final bid to win over opinion, Snow gathered information from across London and prepared a further map; this time comparing mortality rates between the customers of two rival water companies, the Lambeth Water Company and the Southwark and Vauxhall Company, during both the 1849 and 1853–1854 cholera epidemics. In the earlier of these outbreaks, death rates had been similar, but only five years later Southwark and Vauxhall had become almost six times more dangerous a supplier. The difference in 1853 was, as Snow pointed out, that the Lambeth company had moved its supply source to Thames Ditton above the tidal reach of the Thames as required by the 1852 Act. The Southwark and Vauxhall Company had yet to take action. It is for this study that Snow best deserves his recognition as the father of epidemiology, yet it was not until after his death, and further studies resulting from the 1866 outbreak, that medical opinion was generally

convinced. Only in 1870 did John Simon, the City Medical Officer, overcome his adherence to the miasma doctrine to recognise that Snow had been correct. In 1882, the German bacteriologist, Robert Koch successfully identified the waterborne bacillus that caused cholera. With the commissioning of the sewage pumping stations at the end of Bazalgette's drainage system, both typhoid and cholera epidemics were eliminated. 1866 was the last year that cholera took hold in London.

The water supply was taken into public ownership in 1904 and rationalised into one system under MBW. Between 1988 and 1993 an 80 kilometre long water ring main (now the Thames Water Ring Main) was laid 40 metres underground in the London clay—linking treatment works to supply shafts. In 1989 the water authority was privatised again.

Health continues to be monitored by using maps to reveal critical underlying patterns, as demonstrated by the London Health Commission's recent work, whereby they regularly prepare maps to compare health and wellbeing with other social and demographic factors. Snow's achievements, especially his Broad Street map, are now celebrated worldwide. In Broad Street (now Broadwick Street, just behind Oxford Street) visiting celebrants can drink in the John Snow pub on the former site of his surgery. In 2003, the journal *Hospital Doctor* voted him "the greatest doctor of all time", beating Hippocrates into second place.

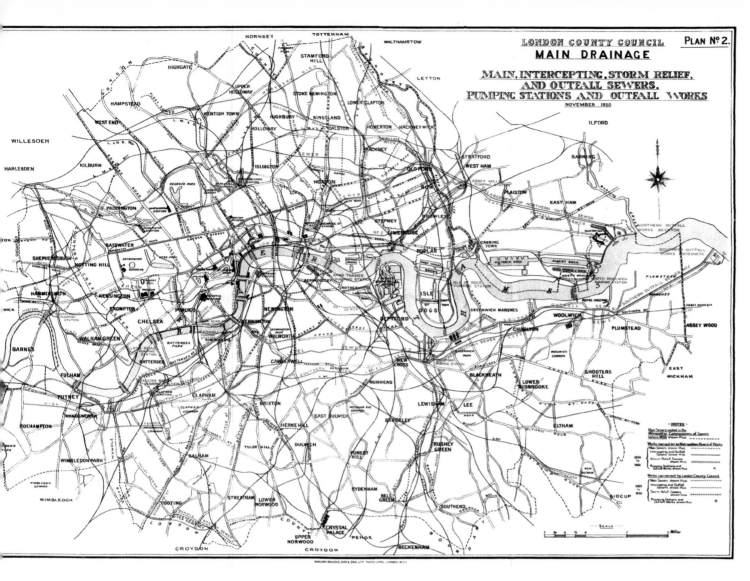

LONDON COUNTY COUNCIL PLAN Nº 2.
MAIN DRAINAGE
MAIN, INTERCEPTING, STORM RELIEF,
AND OUTFALL SEWERS.
PUMPING STATIONS AND OUTFALL WORKS
NOVEMBER 1930

Map of sewerage works in London from
the Main Drainage of London Report,
London County Council, 1930.
Sir George W Humphreys

This 1930s map of Bazalgette's system, shows another
layer of complexity to the drainage layout. The
system is one of the great achievements of Victorian
engineering and was undoubtedly instrumental in
enabling London's rise to predominant world city.
Unlike other European cities, London, through its
effective drainage, managed to stay relatively healthy
and keep major epidemics at bay.

FIRE AND INSURANCE

London has a history of immolation. Boudicca burnt out Roman London in 60 AD and it was destroyed again 60 years later. Stephen Inwood catalogues the "disastrous fires in 1077, 1087 (in which St Paul's and 'the largest and fairest part of the whole city' was destroyed), 1092, 1100, 1133 and 1136", as well as one in Southwark in 1212 and two more in 1220 and 1227.[10] Frequent but lesser fires continued through the centuries, including a major conflagration in 1632, however they all barely rank beside the Great Fire of 1666 that made upwards of 65,000 people homeless.[11] Not until the Second World War, did London burn in a comparable way again, despite the 5,000 fires recorded in the city in a typical period between 1833 and 1841.[12]

The Great Fire, having provoked an overwhelming flurry of mapping in late seventeenth century London, also prompted the first fire brigades and the invention of suitable pumps with 'fire engines' to carry them, as well as a demand for fire insurance. Companies that sprung up to serve the insurance demand included Nicholas Barbon's Fire Office in 1680, the Friendly Society in 1684, the Hand-in-Hand in 1696 and Sun Fire Office in 1710. By the end of the eighteenth century—following the stock market crash caused by the collapse of the South Sea Company in 1720—the fire insurance market had consolidated into the hands of three major companies: The Sun, Royal Exchange and the Phoenix. Richard Horwood compiled his plan of London of 1792–1799 for the Phoenix Fire Office, and dedicated it to their "Trustees and Directors", indicating the importance to such companies of high quality maps and doubtless the financial support to mapmakers that came with it.

These three companies each had their own small fire fighting force that independently dealt with fires such as those in Wapping, in 1716 and 1725, Exchange Alley, 1748, Albion Mills, just over London Bridge, 1791, and the Ratcliff Highway, 1794, as well as tackling the fires caused by the Gordon Riots in 1780. Each of these fires was followed by the preparation of a full plan of the site to allow accurate valuation of any insurance claim, just as had been carried out in the city in 1666. The three separate forces were merged in 1832 along with those of six other smaller companies, to form the London Fire Engine Establishment.

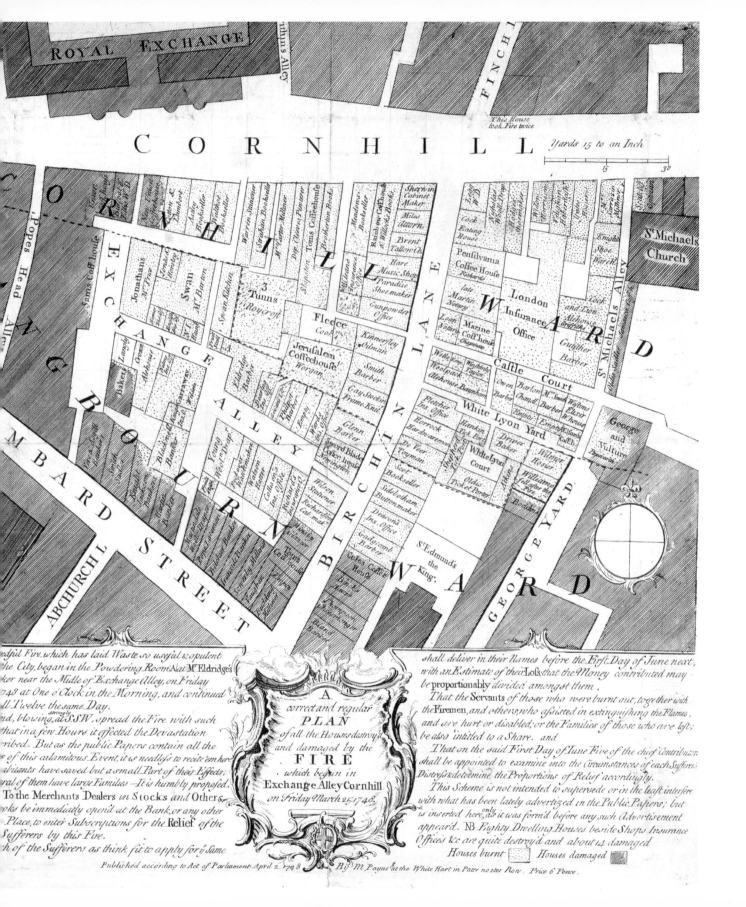

Fire in Exchange Alley, Cornhill, 1748
Thomas Jeffrey

On 25 March 1748 fire broke out at the Royal Exchange Assurance in Exchange Alley in the Cornhill ward of the City. It destroyed the best part of two blocks, including Jonathan's Coffee House, the heart of the emerging stock exchange. Insurers were quick to learn from the event, and other similar disasters in London, and the fire was mapped carefully on their behalf. This was essential to ensure a satisfactory allocation and settlement of any claims, and was also useful, not only informing actuarial calculations, but also in helping to avoid and extinguish such fires in the future.

Image courtesy of Ashley Baynton-Williams.

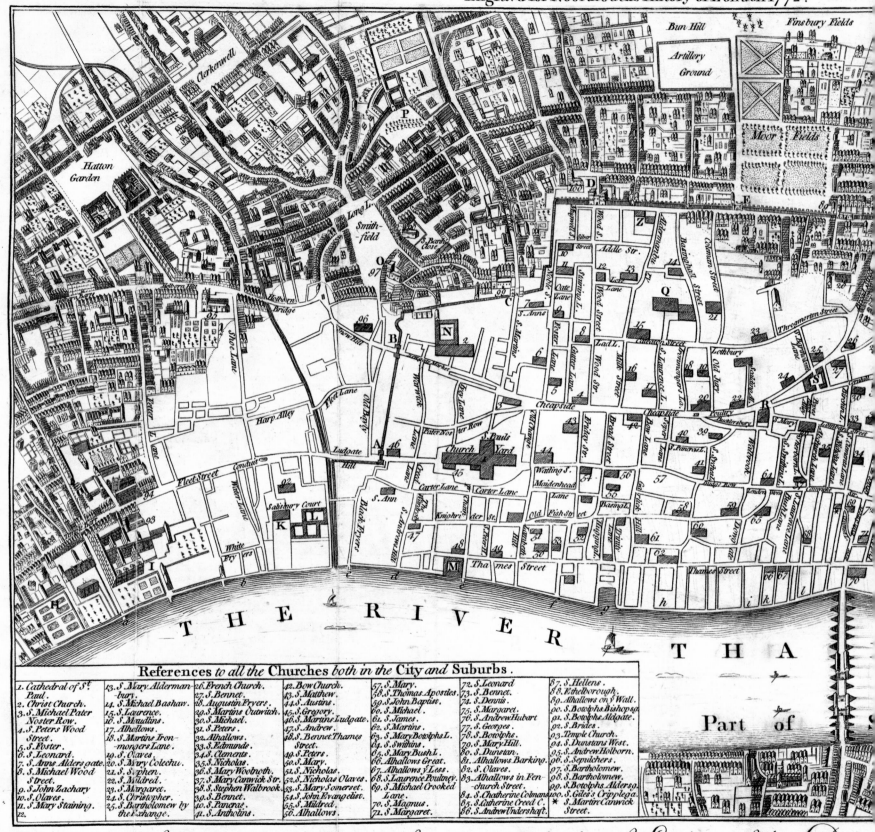

Engrav'd for Noorthouck's History of London 1772.

Finsbury Fields

Bun Hill

Artillery Ground

Moor Fields

Clerkenwell

Hatton Garden

Long Lane

Smith-field

S. Barth Close

Addle Str.

Coleman Street

Threadneedle Street

Holborn Bridge

Shoe Lane

Fetter Lane

Harp Alley

Fleet Street

Conduit

Water Lane

White Fryers

Salisbury Court

Ludgate Hill

Black Fryers

S. Ann

S. Andrew Hill

Knightrider St.

Old Fish Street

Thames Street

Thames Street

Cheapside

Cheapside

S. Pauls Church Yard

Watling St.

Maidenhead Lane

Carter Lane

Pater Noster Row

Wood Street

Aldersgate Street

Lothbury

Poultry

Bucklersbury

S. Mary

Walbrook

THE RIVER THA

Part of

References to all the Churches both in the City and Suburbs.

1. Cathedral of St. Paul.	13. S. Mary Aldermanbury.	26. French Church.	42. Bow Church.	57. S. Mary.	72. S. Leonard	87. S. Hellens.
2. Christ Church.	14. S. Michael Bashaw.	27. S. Bennet.	43. S. Matthew.	58. S. Thomas Apostles.	73. S. Bennet.	88. Ethelborough.
3. S. Michael Pater Noster Row.	15. S. Laurence.	28. Augustin Fryers.	44. S. Austins.	59. S. John Baptist.	74. S. Dennis.	89. Alhallows on y̆ Wall.
4. S. Peters Wood Street.	16. S. Maudlins.	29. S. Martins Outwich.	45. S. Gregory.	60. S. Michael.	75. S. Margaret.	90. S. Botolphs Bishopsg.
5. S. Foster.	17. Alhellows.	30. S. Michael.	45. S. Martins Ludgate.	61. S. James.	76. S. Andrew Hubart.	91. S. Botolphs Aldgate.
6. S. Leonard.	18. S. Martins Ironmongers Lane.	31. S. Peters.	47. S. Andrew.	62. S. Martins.	77. S. Georges.	92. S. Brides.
7. S. Anns Aldersgate.	19. S. Olaves.	32. Alhallows.	48. S. Bennet Thames Street.	63. S. Mary Botolphs L.	78. S. Botolphs.	93. Temple Church.
8. S. Michael Wood Street.	20. S. Bury Colechu.	33. S. Edmunds.	49. S. Peters.	64. S. Swithins.	79. S. Mary Hill.	94. S. Dunstans West.
9. S. John Zachary	21. S. Stephen.	34. S. Clements.	50. S. Mary.	65. S. Mary Bush L.	80. S. Dunstan.	95. S. Andrew Holborn.
10. S. Olaves.	22. S. Mildred.	35. S. Nicholas.	51. S. Nicholas.	66. Alhallows Great.	81. Alhallows Barking.	96. S. Sepulchers.
11. S. Mary Staining.	23. S. Margaret.	36. S. Mary Woolnoth.	52. S. Nicholas Olaves.	67. Alhallows y̆ Less.	82. S. Olaves.	97. S. Bartholomew.
12.	24. S. Christopher.	37. S. Mary Canwick Str.	53. S. Mary Somerset.	68. S. Laurence Poultney.	83. Alhallows in Fenchurch Street.	98. S. Bartholomew.
	25. S. Bartholomew by the Exchange.	38. S. Stephen Walbrook.	54. S. John Evangelist.	69. S. Michael Crooked Lane.	84. S. Chatherine Colmans.	99. S. Botolphs Aldersg.
		39. S. Bennet.	55. S. Mildred.	70. S. Magnus.	85. S. Catherine Creed C.	100. S. Giles's Cripplega.
		40. S. Pancras.	56. Alhallows.	71. S. Margaret.	86. S. Andrew Undershaft.	★ S. Martin Canwick Street.
		41. S. Antholins.				

A PLAN of the CITY and LIBERTIES of LONDON; shewing the Extent of the Drea

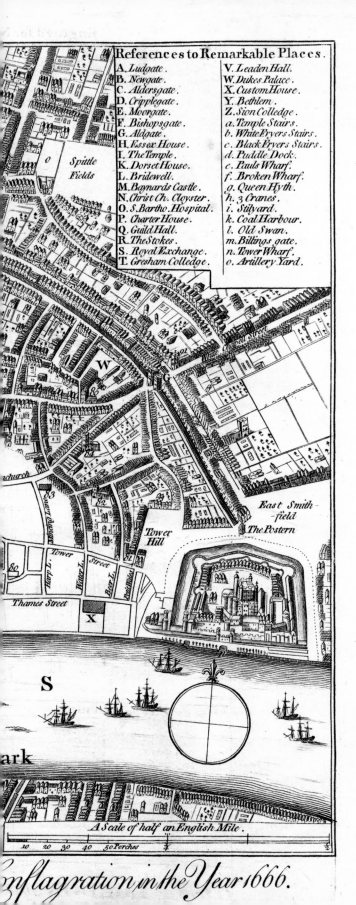

References to Remarkable Places.

A. *Ludgate*.
B. *Newgate*.
C. *Aldersgate*.
D. *Cripplegate*.
E. *Moorgate*.
F. *Bishopsgate*.
G. *Aldgate*.
H. *Essex House*.
I. *The Temple*.
K. *Dorset House*.
L. *Bridewell*.
M. *Baynards Castle*.
N. *Christ Ch. Cloyster*.
O. *S. Bartho. Hospital*.
P. *Charter House*.
Q. *Guild Hall*.
R. *The Stokes*.
S. *Royal Exchange*.
T. *Gresham Colledge*.
V. *Leaden Hall*.
W. *Dukes Palace*.
X. *Custom House*.
Y. *Bethlem*.
Z. *Sion Colledge*.
a. *Temple Stairs*.
b. *White Fryers Stairs*.
c. *Black Fryers Stairs*.
d. *Puddle Dock*.
e. *Pauls Wharf*.
f. *Broken Wharf*.
g. *Queen Hyth*.
h. *3 Cranes*.
i. *Stilyard*.
k. *Coal Harbour*.
l. *Old Swan*.
m. *Billings gate*.
n. *Tower Wharf*.
o. *Artillery Yard*.

Spittle Fields

East Smith-field

The Postern

Tower Hill

Tower Street

Thames Street

A Scale of half an English Mile.

10 20 30 40 50 *Perches*

...nflagration in the Year 1666.

A Plan of the City and Liberties of London, shewing the extent of the dreadful conflagration in the year 1666, 1773
John Noorthouck

The Great Fire of London in 1666 dominates histories of the city, including this plan by Noorthouck, derived from a contemporary map by Robert Pricke, over 100 years after the event. It reminds us that despite the precautions put into place following 1666—such as the two Rebuilding Acts (given royal assent in 1667 and 1670 respectively), which restricted buildings to structures of a maximum of four storeys, with no projecting jetties or windows—that fire was an ever-present threat and danger.

Image courtesy of David Hale/MAPCO.

A plan of the Great Fire in Bishopsgate Street, 1765

The fire that occurred in Bishopsgate Street, just after midnight on 7 November 1765, destroyed 49 houses and shops belonging to tailors, milliners, gunsmiths, wool drapers and cabinet makers alike. Insurance companies were keen to limit their future losses—the preparation of an accurate map, such as this one, was an essential step to achieve this.

Image courtesy of David Hale/MAPCO.

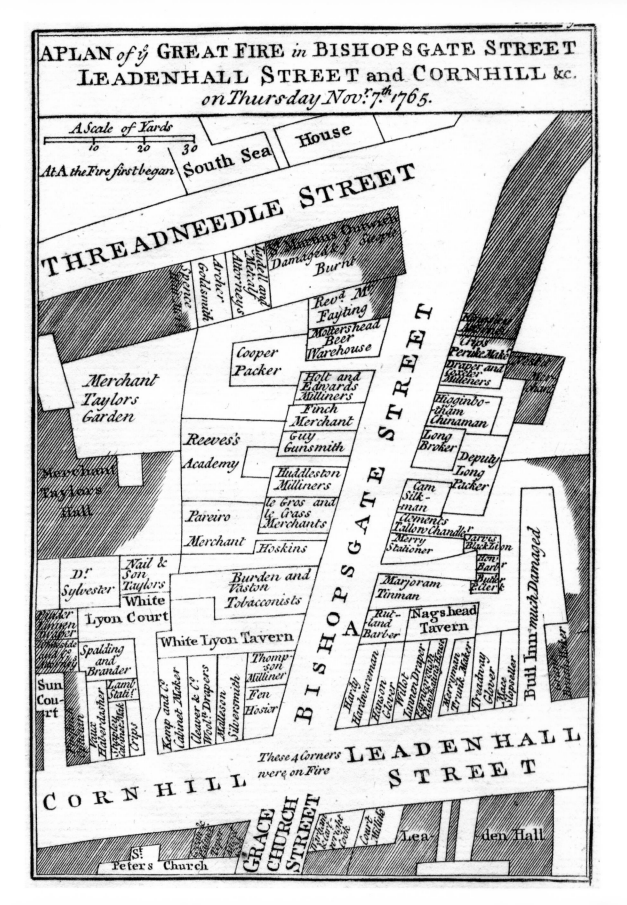

A PLAN of ye GREAT FIRE in BISHOPSGATE STREET LEADENHALL STREET and CORNHILL &c. on Thursday Novr. 7th 1765.

The new brigade embarrassingly failed to deal effectively with major fires at the Palace of Westminster, in 1834, where both Houses of Parliament were destroyed, at The Royal Exchange, the home of London's insurance industry, in 1838 and the warehouses of Tooley Street, again just south of London Bridge, which were destroyed by a two day fire in 1861. As a result, the private system could not last and was replaced in 1865 by the new Metropolitan Fire Brigade under the jurisdiction of the Metropolitan Board of Works (MBW).

The rapidly growing Fire Brigade established new stations across London, as shown on the marked-up version of Harris's plan of 1791 in the collections of the British Library. The insurance industry, having lost the ability to control fires once they had broken out, began to develop new ways of using maps to establish risks—therefore premiums in advance of any damage—with Charles Goad publishing his first fire insurance maps in 1885. The maps give detailed assessments of areas and buildings specifically for insurance purposes, but they are also some of the most accurate local maps available, and were rigorously kept up to date by the use of overlays—making them an exceptional resource for historians. Goad's company continued to produce such maps, concentrating on retail and commercial property until 1970.

In today's world, the London Fire Brigade uses Geographical Information Systems (GIS) and thematic mapping techniques to help plan its service and to tackle problems such as malicious false alarm calls. The information provided by new mapping technologies is beginning to transform the work of the emergency services, helping them to predict and to reach incidents with the appropriate equipment and manpower as early as possible.

Goad Fire Insurance Plan of the City of London
(showing the area between Mark Lane, Billiter
Street, Leadenhall Street, Northumberland Alley
and Crutched Friars), 1887
Charles Goad

By the late nineteenth century, plans began to be
prepared in advance of potential fires—in order to
aid insurance companies in assessing fire risk, the
likelihood and extent of possible damage and, of
course, the premium that the property owner would be
charged. The most significant producer of these maps
was the firm of Charles Goad, which mapped cities,
towns and villages across the world.

The maps Goad created provide a detailed record
of London throughout the years, and were produced
and regularly revised. Such maps included street
widths, property numbers, occupiers and their trades,
and a colour-coding system that classified different
construction types: red for brick buildings, yellow for
timber and dark blue for stone.

Image courtesy of the Guildhall Library, City of London.

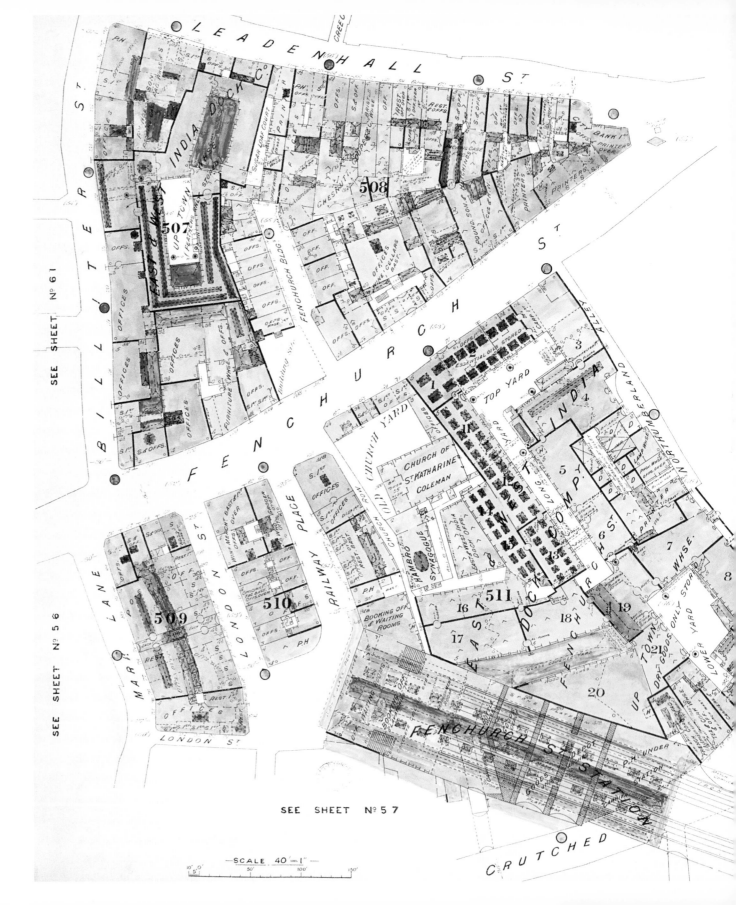

Location of Fire Brigade Stations, circa 1880

The establishment of the Metropolitan Fire Brigade, funded by public money rather than insurance companies, transformed fire fighting in the capital. Ex-sailors were recruited as full-time firefighters and manned the 55 main stations across the MBW area of London. This map, from a MBW atlas, shows the location of such stations, in positions from which they could reach fires quickly and effectively.

Image courtesy of Thames Water.

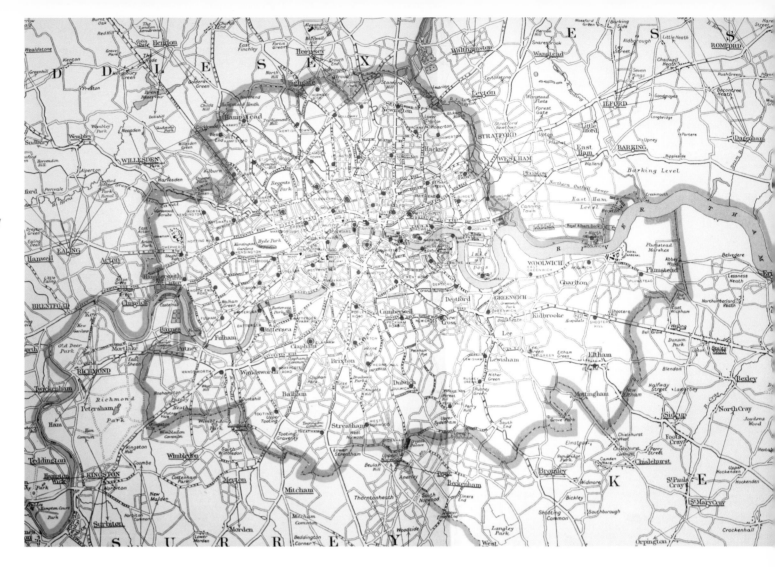

UTILITIES AND SERVICES

On New Year's Eve, 1813, the lights on Westminster Bridge were lit for the first time with gas by the newly established London and Westminster Gas Light and Coke Company (GLCC). Nearly 70 years later, another bridge, the Holborn Viaduct, was lit by electric light. Neither provided much brightness, but each was the start of a new service that would transform London. Nineteenth century London was not only growing at a pace that meant its population increased by approximately 20 per cent every ten years—lifting it from 958,863 in 1801 to 6,586,000 in 1901—it was also installing unprecedented new services and utilities that made such growth and vigour possible. In addition to receiving gas and electricity London had, from 1883, a distribution system for hydraulic power covering an area from Hyde Park to Vauxhall. Rowland Hill's Penny Post had transformed the postal delivery system from 1840 on and the first commercial telegraphs were sent in 1837.

With the exception of the postal system, these services were provided by a range of competing private companies and, in the case of the electrical supply, the 1882 Electric Lighting Act actively encouraged large numbers of small companies to enter the market (over 100 companies emerged in 1882, with many of them rapidly going bankrupt thereafter). A harsh Darwinian experiment seemed to be at play at the birth of each industry—whether water, gas or electricity—with only the fittest surviving in each case.

The GLCC, the first gas company to be established, was by the 1850s, only one of 20 companies competing to offer street lighting to wards, parishes and private estates. In 1857, the companies abandoned competition and divided London into 13 monopoly areas, simultaneously raising their charges. The Metropolitan Board of Works (MBW), having failed to achieve a municipal gas monopoly as in other English cities such as Birmingham and Bradford, managed to garner some control over the quality of the supply, if not the price, as it had wanted. The GLCC was nationalised in 1949, along with 12 other companies, to become the North Thames Gas Board. The Board lasted until 1973 when it was subsumed into British Gas.

The electricity supply business attracted similarly large numbers of hopeful entrepreneurs at its outset, although the initial problems of competing with the well-established and organised gas business quickly outstripped many of the fledgling companies—and a boom and bust in electricity shares in the stock market in 1822 almost collapsed the industry before it had a chance to find its feet.[13]

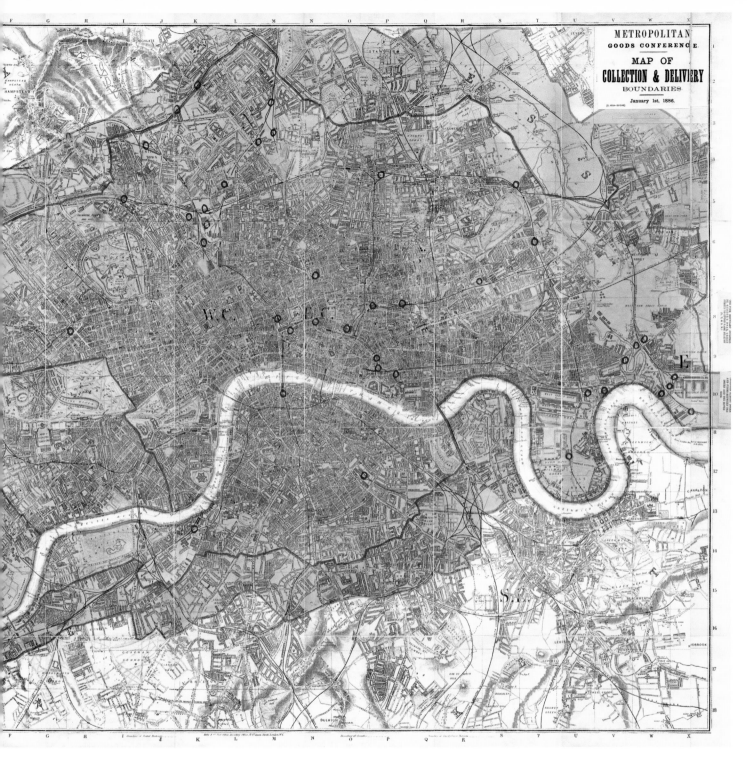

Map of collection and delivery boundaries, 1886
Kelly & Co. PO Directory Office

Kelly & Co. produced maps for the Post Office from
1840 onward. This map shows the boundaries—all
brightly and fairly aggressively coloured—for the
delivery and collection of packages and goods
across London in 1886.

Image courtesy of David Hale/MAPCO.

SKETCH MAP OF DISTRICT
WESTMINSTER ELECTRIC SUPPLY CORPTN. LTD

Generating Stations of the Westminster Electric Supply Corporation Ltd from *Popular Electric Lighting* by E Ironside Bax, 1891

London had a multitude of electricity companies at the end of the nineteenth century, all competing to run DC mains supplies down whichever streets they could. This map shows just some of the areas that the Westminster Electric Supply Corporation provided for. (They also distributed electricity to the areas of Belgravia and Mayfair.)

Image courtesy of the Science & Society Picture Library.

London Electrical Supply from Bacon's *Atlas
of London*, circa 1900
George Bacon

Until nationalisation in 1947, London was supplied by
a multiplicity of electrical companies. As shown on
this map, each company had its own monopoly area
in London. However, the small-scale of most of the
operators resulted in inefficient and expensive supplies,

and local interests fought off any attempts at unification
for far longer than any of London's, or Britain's,
commercial rivals.

Image courtesy of Thames Water.

Map shewing the several walks or deliveries in the
Country Districts of the Two-Penny Post, 1830
Aaron Arrowsmith

At a time when the British postal system was in urgent
need of reform, and just before the introduction of
the Penny Post in 1840, this map was commissioned
by the House of Commons for a report of the
Commissioners of Revenue Inquiry and shows the area
of each of the delivery and collection walks or rides as
operated by the Two-Penny Post system.

Image courtesy of the Guildhall Library, City of London.

An 1891 map of the Westminster area shows the distribution network of the Westminster Electric Supply Corporation Ltd, one of the companies that survived the initial rout to set up a flourishing business and develop a close working relationship with the Grosvenor Estate.

Maps were clearly necessary for companies to plan such services, but prior to the universal Penny Post service, customers also needed maps to calculate the cost (usually to the recipient) of mail delivery. Maps were published to show the limits of delivery walks and allow calculation of prices for sending letters or packages outside the central area. From 1711–1801, post in London, and within a radius of ten miles from the central post office, cost one penny (1d) per ounce. After 1801, and up until the introduction of the universal penny post, this was raised to 2d. Letters to the suburbs cost 3d after 1801 and usually 4d beyond that. The maps also show the fixed 'post walks' that the 224 urban and 165 suburban letter carriers walked six times each day—three times to collect and three more to deliver the mail. Such a system made it possible to write a letter and receive a reply on the same day—a service only achievable with private couriers at much greater cost today.

The uniform Penny Post revolutionised this labourious approach, which saw the carriers deliver in the region of only 200 letters a day, despite putting in close to 11 hour days. Almost immediately, the amount of post more than doubled and the service has not looked back since, although it would take many years before returning to profit. Unfortunately, having a single rate made the many maps that explained the complex system of charging for the distance covered redundant. Not until the idea of zoning London for transport ticketing would such maps appear again.

As in the early nineteenth century, the late twentieth saw a new range of services rapidly and widely become available to Londoners; from motorcycle couriers, to mobile phones, WiFi and satellite navigation. New services require new maps and, sometimes, new forms of mapping. Maps are available that show the strengths of various communications signals as well as the systems architecture of city-wide installations. Maps showing the services and utilities available to Londoners have never been more available, more complex to understand or more elaborate in their design.

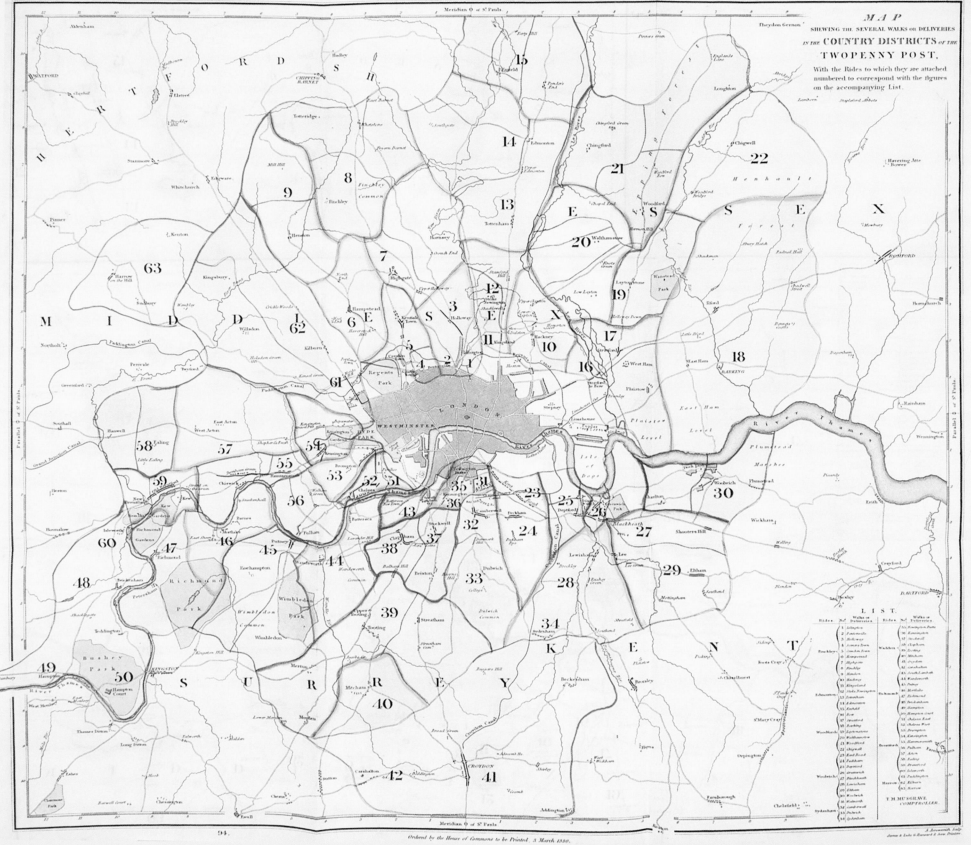

MAP
SHEWING THE SEVERAL WALKS OR DELIVERIES
IN THE COUNTRY DISTRICTS OF THE
TWOPENNY POST,
With the Rides to which they are attached
numbered to correspond with the figures
on the accompanying List.

Cary's new Plan of London and its vicinity shewing
the limits of the Two-Penny Post delivery, 1837
John Cary

This map depicts the full extent of the Two-Penny
Post delivery. Published for the Postmaster General,
it also shows the number of different services in
London, including transport routes and various
administrative boundaries.

Image courtesy of David Hale/MAPCO.

PLANNING THE CITY

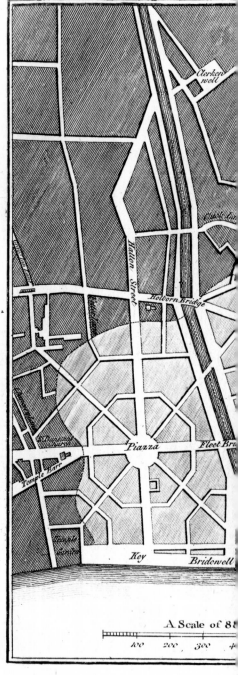

In London, the great unplanned city, where areas of development overlap and co-exist to create a patchwork of patterns and places; planning has always arrived in concentrated waves, anxious to improve the disreputable old city and find a better way to organise it. Such attempts to tame London's unruly nature have been triggered by disasters, such as the Great Fire of 1666 or the bombing of London from 1939–1945, or by the perceived need to expand and modernise; to create new towns or a new financial centre in the Docklands. The city has only ever responded with the greatest reluctance to visionaries with far-reaching plans for reform. London had no Hippodamus (the fifth century Greek planner) Hausmann (the nineteenth century city planner of Paris) or Cerdà (the nineteenth century city planner of Barcelona). The pragmatic pursuit of profit has changed and shaped London more than any other comparable force—but that hasn't stopped a long series of luminaries pursuing and promoting their own alternative visions for the city.

The plans developed by Wren, Nash, Abercrombie, or within the Greater London Authority (GLA), are maps of an alternative London. Alternatives that might have been, and may yet be, lived in and visited by future generations. They are also plans that consolidate a mass of policy goals; imagined realities and the rebuilt Jerusalem of Blake's poetic visions.

The plan of the future is less precise than the mapping of the present, although the roads are straighter and activities will apparently be confined to neatly radiating circles. The mapped future is always cleaner and tidier. It is also brighter and bolder and will be best appreciated from high above.

The temptation to redesign the map of London has always been strong. Its streets are frustratingly misaligned and many of its public spaces are pinched and mean. John Gwynn, architect, founding member of the Royal Academy and close friend of Samuel Johnson, published *London and Westminster Improved* in 1776—an intelligent project for restructuring the streets of London to create through-routes and rational street patterns, as well as vista and prospects in the city. He noted that:

> Custom has hitherto blinded the inhabitants of London,
> with respect to the notorious inconveniences, and the popular
> prejudice is deeply rooted in them that London is, in every
> respect, the finest city in the world, prevents the majority from
> seeing and considering its defects and consequently they quietly
> submit to be thrust more than half a mile out of the way rather
> than call into question the understanding of their forefathers.[14]

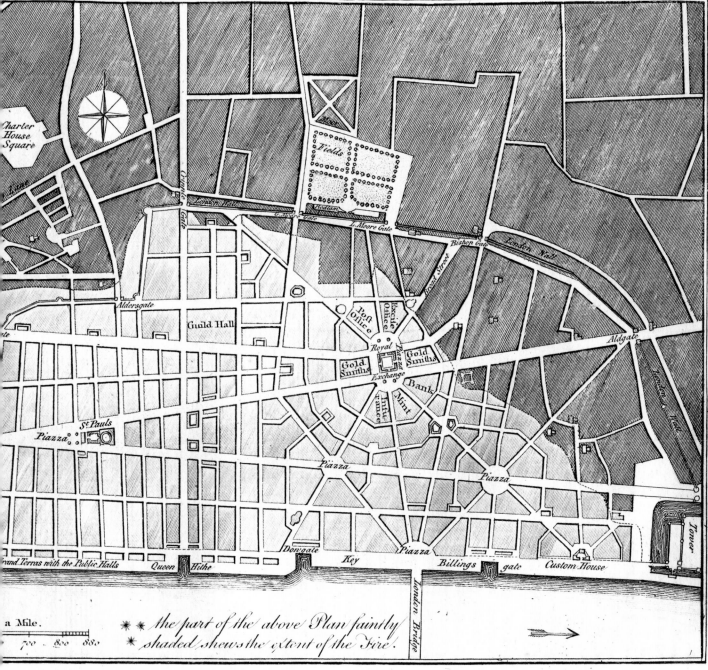

Engraved for Harrison's History of London.

Charter
House
Square

Fields

Aldersgate

Guild Hall

Post
Office

Excise
Office

Royal
Gold
Smiths

Gold
Smiths

Exchange

Bank

Mint

St Pauls

Piazza

Piazza

Piazza

Piazza

Bishop Gate

Aldgate

Tower

a Mile.
700   800   880

Grand Terras with the Public Halls   Queen Hithe   Dowgate   Key   Billings gate   Custom House

London Bridge

** ★ the part of the above Plan faintly
    ★ shaded shews the extent of the Fire.

Plan for Rebuilding the City of London after the dreadfull Conflagration in 1666.

Plan for the rebuilding of the City of London
after the Great Fire in 1666, 1666
Christopher Wren

Wren saw the Great Fire as the perfect opportunity
to show off his—as yet—underdeveloped talents.
Comprised of radiating streets that met at rond-points,
and rational grids of lanes, his plan for London ignored
existing road alignments and any sensitivities about
land ownerships. It would have been an astonishing
achievement had it been realised, however, like
all the other plans submitted, it was doomed to
failure—a pragmatic solution based on existing land
ownership, and that could achieve the required speed
of reconstruction, was chosen instead. After this
disappointment, Wren took on the job of the King's
Surveyor of Works, which included the task of rebuilding
St Paul's Cathedral and the many City churches.

Image courtesy of David Hale/MAPCO.

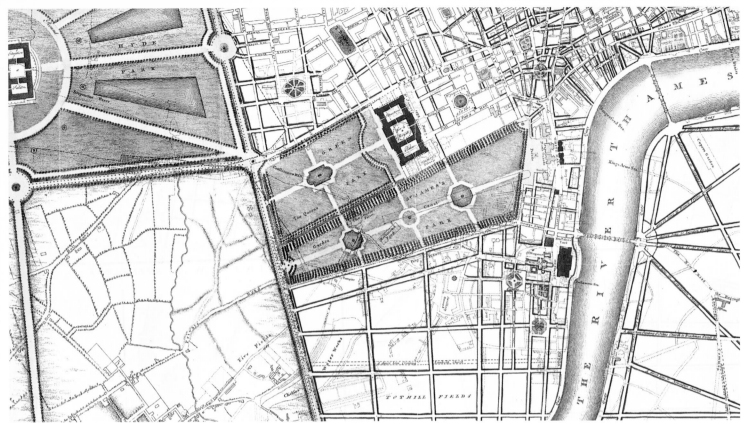

Hyde Park: *London and Westminster*
Improved, 1766
John Gwynn

John Gwynn, architect, engineer, founder member of the
Royal Academy and good friend of Samuel Johnson, was
convinced that London was in need of rational planning.
He revisited Christopher Wren's plan of 1666 for the
City—post-Great Fire—and prepared his own proposals,
in which he straightened out West End roads and
created new attractions and landmarks in suitably grand
locations. His plan for Hyde Park included an immense
royal palace with a grand avenue leading to St Paul's.

Gwynn's approach to mapping London was one of
invasive surgery, designed to tidy up its eccentricities
and provide new parts only when necessary. His was a
compelling view of how to re-plan the city.

Image courtesy of the Guildhall Library, City of London.

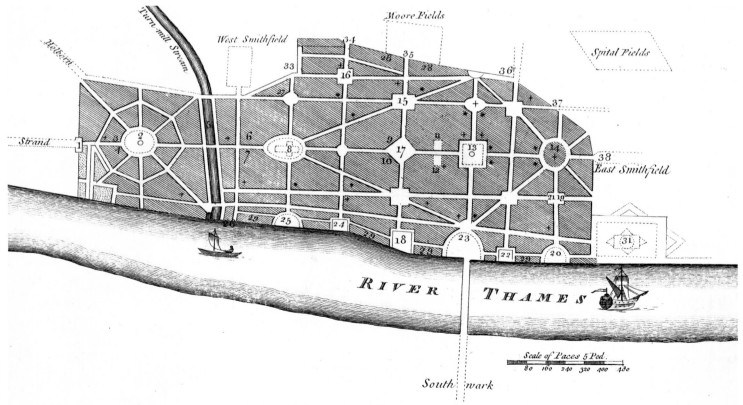

The rest of the openings are for the Markets &c. And in the intermedial
Squares and Areas, what narrower Streets shall be thought fit :

Plan for rebuilding after the Great Fire, 1666
John Evelyn

Evelyn, best known for his diaries rather than his
skills as a town planner, developed a new city plan
perfectly composed of axes, views, small piazzas and
landmark sculptures.

This plan was abstract, rather than a serious
proposition, and barely concerned with the areas beyond
those devastated by the fire—with only the Thames and
the position of London Bridge hinting at these.

Image courtesy of David Hale/MAPCO.

Gwynn proposed a new, large royal palace for the centre of Hyde Park and that the Houses of Parliament, including Westminster Hall, be demolished (anticipating and relishing the outcry that this would cause) in order to replace it with something better. He wanted to create space, as demonstrated by one of his proposals that "a large street is to be opened running south to the Thames".[15] He believed in the grand gesture and clearly thought that London needed many more of them.

Following in Gwynn's footsteps, the redoubtable Lieutenant Colonel Frederick William Trench MP took the notion of the grand gesture a good deal further. In 1825, he published proposals (prepared by the architect brothers Benjamin and Phil Wyatt) for an even more impressive palace in the park, and connected it to St Paul's Cathedral with a two mile long triumphal route carving its way straight through the existing street pattern of London. A contemporary, Mrs Arbuthnot, who saw the drawings in 1825 commented in her diary: "It is the most ridiculous plan I ever saw for, added to it, is the idea of a street 200 feet wide extending from the end of Hyde Park opposite the New Palace to St Paul's! The King and the Duke of York are madly eager for the plan; but the former says he supposes his dammed ministers won't allow it."[16]

Gwynn and Trench may have only wanted to correct a few of the deficiencies of London, but Willey Reveley published proposals in 1796 to deal with nature's flawed contribution to London's layout and to straighten out the route of the Thames itself, through Ratcliffe, the Isle of Dogs and Blackwall—creating, as a result, three docks in the leftover bends of the river.

The first serious outbreak of re-planning fever followed the Great Fire in 1666. On 11 September, only a week after the last embers had been doused, Christopher Wren presented a proposal for rebuilding the city to the King. John Evelyn followed with a different version two days later and on 19 September, having possibly been stung into action by Wren's presentation to the Royal Society the week before, Robert Hooke submitted his own plan to the City Corporation, just in time for a second Royal Society meeting. Peter Mills, the city's surveyor, presented another proposal to the City Corporation and two plans by Richard Newcourt and one by Valentine Knight followed shortly after.

All the competing parties believed that London was ready for something more formal and grand than the tight weave of pre-fire streets. Wren's plan consisted of radial avenues overlaid on a basic

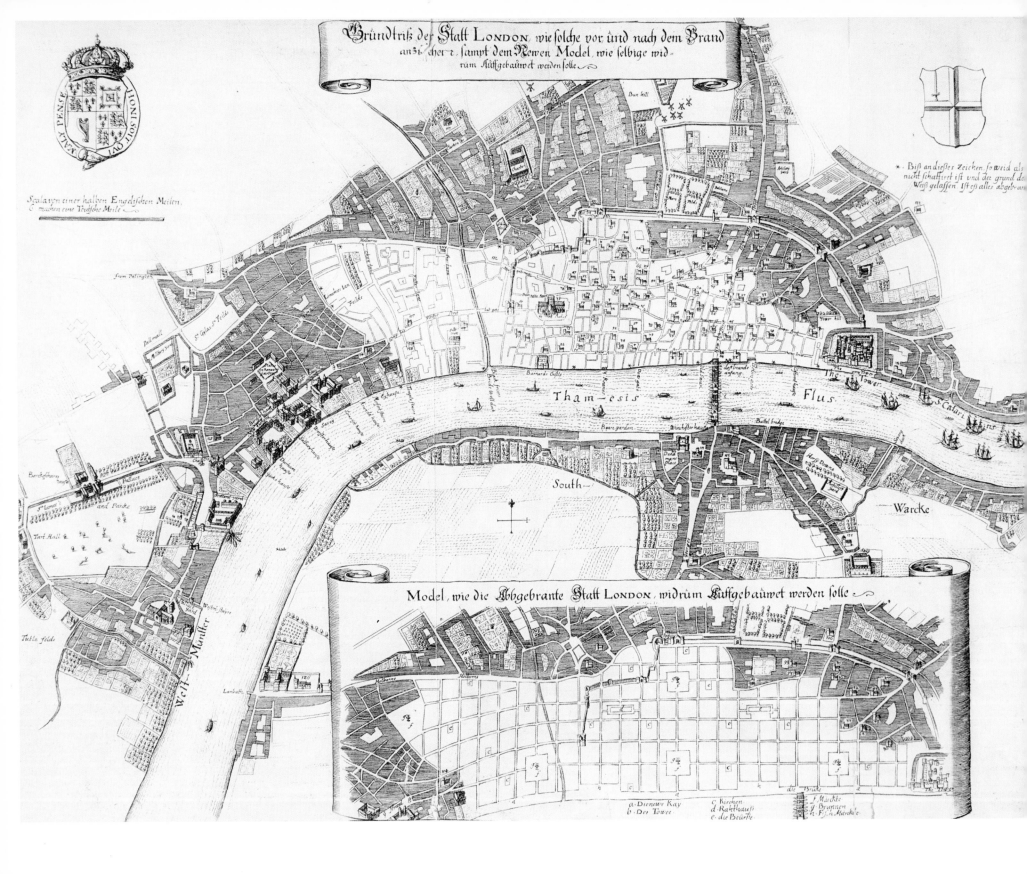

Grundtriß der Statt LONDON, wie solche vor und nach dem Brand anzieher 2 sampt dem Newen Model, wie selbige widrum Auffgebauwet werden solle

Model, wie die Abgebrante Statt LONDON, widrum Auffgebauwet werden solle

Tham-esis   Flus

South—   Warcke

gridiron of streets. St Paul's sits prominently in the angle between the main avenues. He created several rond-points with six or eight streets emerging from octagonal piazzas. Evelyn had a similar approach, with several locations for fountains, but did without the grand avenues. In his plan, St Paul's is placed at the centre of an elliptical piazza. Hooke and Newcourt both went for the regular grid, forming blocks of buildings or squares. And Knight? He divided the area up into long thin plots of land, linked by the minimum number of narrow streets, all designed for profitable sale onto developers and ultimately a profit to the crown. He added to this a strange proposal to run a canal around the city, linking the Thames to the Fleet River, but for his pains was arrested for suggesting that the king "would draw a benefit to himself, from so publick a Calamity of his people".[17] The scheme died a death, as did they all; neither the Corporation nor the crown had the stomach or the finances for re-planning the area. The old street pattern was retained, while minor improvements and rebuilding went ahead on the old plots—although this time in stone and brick rather than timber.

Wren was rewarded with the commission for the building of City churches as well as the new cathedral but his Baroque plan for London, although never implemented, was never entirely forgotten either. John Gwynn reworked and republished Wren's plan in 1849 and many studies have been drawn up since, based on his proposals. London undoubtedly lost the opportunity to redefine a part of its fabric and to create a grand streetscape, but it wasn't about to change the habit of centuries. As Peter Ackroyd suggests: "The very nature of the city defeated them: its ancient foundations lie deeper than the level at which any fire might touch, and the spirit of place remained unscathed."[18]

That spirit has continued to battle with the re-planners of London. John Nash, architect to the Prince Regent, later George IV, managed to create a 'Royal Way' between the new Regent's Park and St James' Park. Even so, however, it twisted and turned, writhing through the existing streets to create Portland Place, Regent's Street and, finally, Waterloo Place before reaching the Mall. Nash's grand vision, despite some wonderfully creative solutions to the turns en route, including the hinge-like entrance tower of All Souls, Langham Place, became almost forgotten over the years and parts of his creation, particularly in the curve of Regent's Street as it approaches Piccadilly Circus, were destroyed. It is now in the process of being

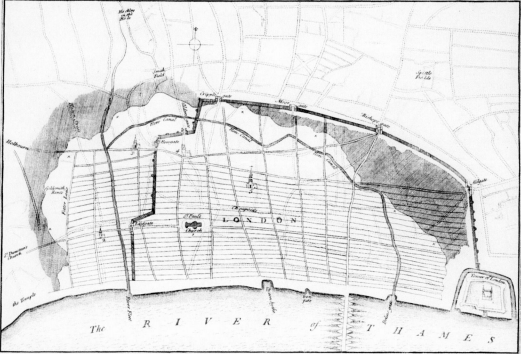

Plan for rebuilding after the Great Fire, 1666
Valentine Knight

Valentine Knight's proposal was treated more seriously than those of his contemporaries and resulted in arrest on order of the King. His plan included a canal, approximately 27 metres wide, which skirted around the effected area (thus creating the opportunity to raise tax revenue from the additional wharves and docksides that would pay for the rebuilding). Charles II was not pleased with what he perceived to be an opportunistic approach on his behalf, and incarcerated Knight.

Otherwise, Knight's proposal is a strange grid of narrow and elongated blocks, accommodating two rows of houses between parallel lanes and occasional cross streets. He worked out his proposals in some detail, including costs and quantities of building materials, as well as the ingenious means to raise finance.

Image courtesy of the Guildhall Library, City of London.

restored and re-established under the direction of Sir Terry Farrell with the rubric of the Nash Ramblas.

London in the Victorian and Edwardian periods was too busy expanding outward to be concerned much with city planning. Terraces of houses, and yet more houses, were laid out as the city spread further and further from its centre. There was almost no grand vision during this expansion. The major schemes that were undertaken by the Metropolitan Board of Works (MBW), particularly the Albert, Chelsea and Victoria Embankments, were frequently the result of another variety of grand project, such as Bazalgette's great drainage, sewage or transport works aimed at making London's traffic flow better. Only the creation of Kingsway and the Aldwych, involving the clearing of 28 acres of slums, workshops and alleyways from 1900–1905, were undertaken for the sake of glorifying the capital and imbuing it with some sense of imperial grandeur.

If there was one visionary who did attempt to change perceptions, it was the Scottish botanist, John Claudius Loudon, who later became a landscape architect (a term he brought into general usage) and, at least in his own estimation, a city planner. He had a long lasting impact on landscape, particularly cemetery design, and published many books and articles on a variety of botanical and landscape topics. It was as a writer and theorist that he tackled the re-planning of London, publishing an article entitled "Observations on Laying out the Public Spaces in London" in 1802 and "Hints on Breathing Places for the Metropolis, and for Country Towns and Villages, on fixed Principles" in 1829. In the latter, Loudon included an idealised plan of London as a series of concentric circles, including the idea of Green Belts almost 60 years before it was proposed by Ebenezer Howard.

The expansion of London into the suburbs, and particularly the arrival of underground trains in outlying areas, did result in one significant exception: the creation of Bedford Park from 1875 onward, managed by the developer Jonathan Carr and based on the designs of, amongst others, Norman Shaw. This early example of a 'garden suburb' led directly to the propositions of Ebenezer Howard in *To-Morrow: a Peaceful Path to Real Reform*, 1898, (reissued in 1902 as *Garden Cities of To-Morrow*) and to the establishment of the Garden City Association in 1899. Howard's familiar town planning diagrams, including The Three Magnets, are his propositions for the future of living in cities such as London, with distinct echoes of Loudon's diagram from over half a decade before.

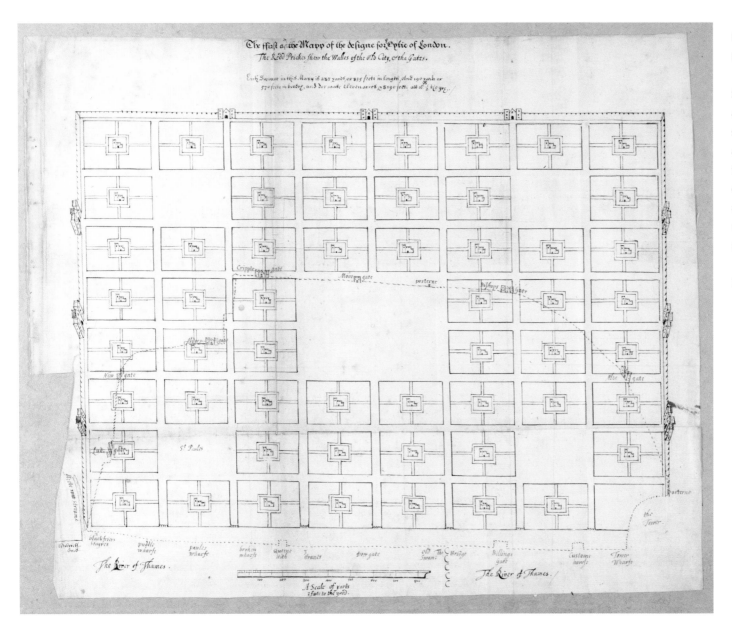

The ffirst of the Mapp of the designe for Citie of London.
The Red Prickes shew the Walles of the Old City, & the Gates.

Each Square in this Mapp is 225 yards, or 675 feet in length, and 190 yards or 570 feet in breadtz, and doc make eleven acres & 3190 feet, att 16 3/4 to yards.

Plan for rebuilding after the Great Fire, 1666
Richard Newcourt

So orthogonal are his streets, and the outer walled and gated edges of the city quarter, that Newcourt's scheme for London is an unvarying one that does not appear to recognise the site it was notionally designed for. Newcourt, as in his pervious maps, seems intent on creating a symmetrical and perfect city and took advantage of the excitement generated over London's redevelopment to reintroduce this concept to the public once again.

Image courtesy of the Guildhall Library, City of London.

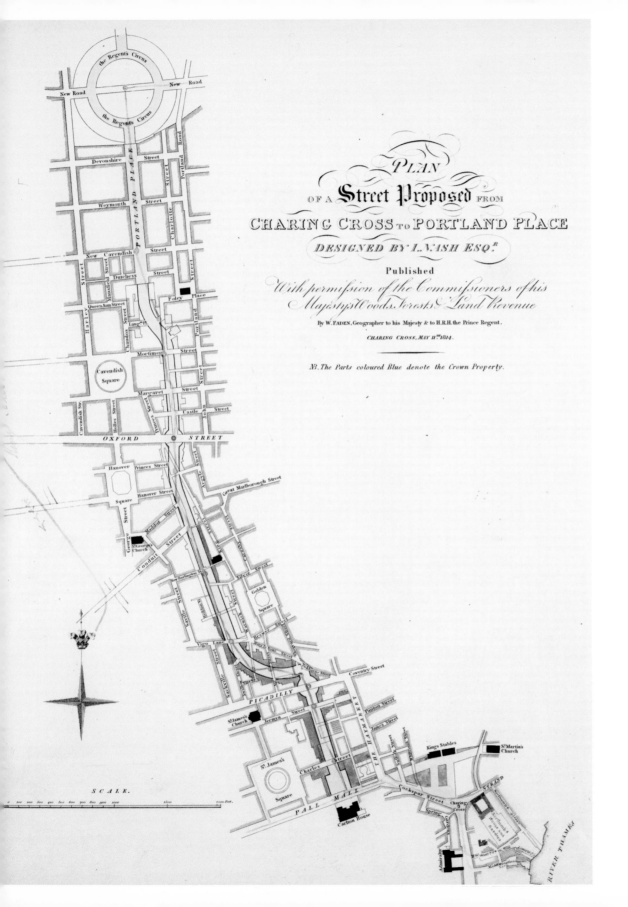

PLAN

OF A Street Proposed FROM

CHARING CROSS TO PORTLAND PLACE

DESIGNED BY I. NASH ESQ.R

Published

With permission of the Commissioners of his
Majesty's Woods, Forests & Land Revenue

By W. FADEN, Geographer to his Majesty & to H.R.H. the Prince Regent.

CHARING CROSS, MAY 11TH 1814.

N.B. The Parts coloured Blue denote the Crown Property.

Plan of a Street proposed from Charing Cross to
Portland Place, 1814
John Nash

In London and Westminster Improved, 1766, John
Gwynne commented on the lack of city planning, and
proposed that a street should run from the New Road
(opened in 1756) into the centre of Westminster.
(The idea had the practical purpose of reducing the
journey time from Westminster to the City, even if
it was by a circuitous route, as well as increasing
the value of the surrounding estates.) In 1811, when
Portland Place reverted to the Crown's ownership,
this became a practical possibility and the House of
Commons commissioned John Nash to design the new
Marylebone Park (now Regent's Park) and a grand
street that ran from Carlton House, overlooking the
Mall, through to the Circus already being created at
Piccadilly and up to Portland Place.

Nash produced a number of variations of the plan.
In them, he had to adapt the road to all number of
twists and turns, like the circular drum and cylinder
entrance tower of All Souls Langham Place and the
great crescent at the bottom of Regent's Street as it
approaches Piccadilly Circus.

The garden city concept included the proposal that settlements should be protected by Green Belts, separating them from adjacent developments and allowing their residents relatively easy access to open space.

Bedford Park was followed by Hampstead Garden Suburb, founded in 1907 by Henrietta Barnett, and planned by Barry Parker and Raymond Unwin. The concept of the low density residential area subsequently gets repeated—without Parker and Unwin's design flair—in suburbs across the country. However, with Hampstead Garden Suburb, the idea of modern town planning had been discovered and, with it, the making of elaborate plans at all sorts of scales and for very different ventures. One of the collaborators on some principal buildings in the suburb, Edwin Lutyens, took to the idea of planning in a serious way (he was appointed to work on New Delhi in 1912). In the last years of his life, together with Sir Charles Bressey, he headed a Royal Academy Committee for the re-planning of London after the Second World War. The Committee concentrated particularly on the road network and fed directly into the two major plans of the twentieth century—*The County of London Plan*, 1943, by JH Forshaw and Patrick Abercrombie and the *Greater London Plan*, 1944, by Abercrombie—notionally on his own.

Like the planning undertaken in 1666, re-planning London after the Second World War, and the destruction caused during the Blitz, was both a necessity and a huge opportunity. The London Country Council (LCC) area, the subject of the first of the great wartime plans, included far too little of the real extent of London for efficient land-use planning and the second *Greater London Plan* was almost immediately commissioned to rectify the problem. The graphics of the best illustrations and maps in both plans are inspiring and convincing in themselves. They express huge confidence in a restored and resurgent London despite the desperate condition of the city when they were being prepared.

The, not unfamiliar, concept of four rings was proposed and London was to be reshaped to fit the diagram better. No new industry was to be allowed in the inner urban zone, and its density was to be capped; requiring the decentralising of 415,000 people. The second zone, the suburban ring, "with regard to population and industry, is to be regarded as a static zone, it is neither a reception area for decentralised persons, nor industry; nor does it in general, require decentralisation".[19] The third zone was effectively the Green Belt, already drawn tight around the Greater Metropolitan Area, and a result of the campaign work of Parker and Unwin and the 1938 Green Belt Act. Any development or expansion of existing communities within this zone was to be strictly limited bar a few exceptions. The fourth ring—the outer country ring—is the receiving ground for the decentralised (both people and industry) and, in time, would include all the new towns to be built around London. It was clearly intended to take the map of chaotic, anarchic London in hand and make it behave more like the proper diagram that its guardians saw as its destiny.

Others, on scrutinising the map, came to different conclusions. Ralph Tubbs, in *Living in Cities*, perceived London as a living organism, albeit a sick one, and planned arteries running into the centre in the form of green leisure corridors.[20] He conceived of a town plan that more closely resembled a flower or snow crystal than the concentric rings of Abercrombie. Two other architects, Kenneth Lindy and Winton Lewis, who jointly published their ideas in 1944, saw the city much as Gwynne had, and believed that the damage resulting from the war proffered the opportunity to create broad streets terminating at significant 'points of focus' such as St Paul's and Guildhall. As in Evelyn's post-fire plan, St Paul's was to be located in an otherwise empty ellipse—but unlike Evelyn, Lindy and Lewis also proposed a monumental, cruciform

heliport over Liverpool Street Station, spanning between five tower blocks. Plans of the post-war years, having addressed wholesale re-planning, population relocation and slum clearance, had built up a head of steam for dealing with the big issues, none of which were bigger than the need to provide enough housing, and to improve traffic flows, through the city.

The impact of housing policy on the character of London was profound. Although developed incrementally, the grain of streets disappeared from patches all over the city. It is as easy to recognise mid- to late twentieth century housing on any map from that of previous eras, whether Medieval, Georgian or Victorian. Point and slab blocks in a landscape of grass and car parking have a very different ground figure even from Edwardian mansion blocks. That estates were themselves planned at a large-scale also tends to reveal the geometric and aesthetic sensibilities of the designers, whether Beaux Arts, Brutalist or in the organic tradition. Often, the specific influences of various overseas architects and design trends can be spotted, just as can the trademark styles of different system-building techniques and the working diameters of the tower cranes used to build the high-rise elements of estates.

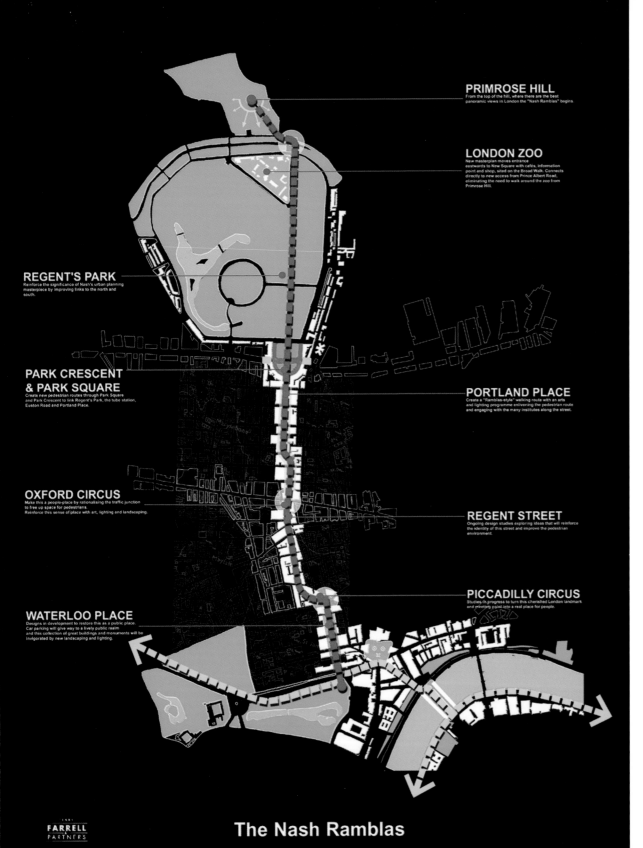

The Nash Ramblas

Nash Ramblas, 2002
Terry Farrell, Farrell & Partners

By the start of the twenty-first century, John Nash's route, which ran from the top of Regent's Park down to the Mall via the Wellington Steps, had become distinctly shabby. Terry Farrell produced proposals for rejuvenating the course, which included opening up the direct road into Regent's Park from Portland Place, evident on Nash's original plan, and exploiting the impact of London's Congestion Zone on parking capacity, by creating a broad pedestrian route down the centre of Portland Place into the heart of the West End (hence the modish reference to Barcelona's famous Ramblas). Some renovation work is already underway on the lower part of the route but the transformation of the upper section is yet to be realised.

Image courtesy of Farrell & Partners.

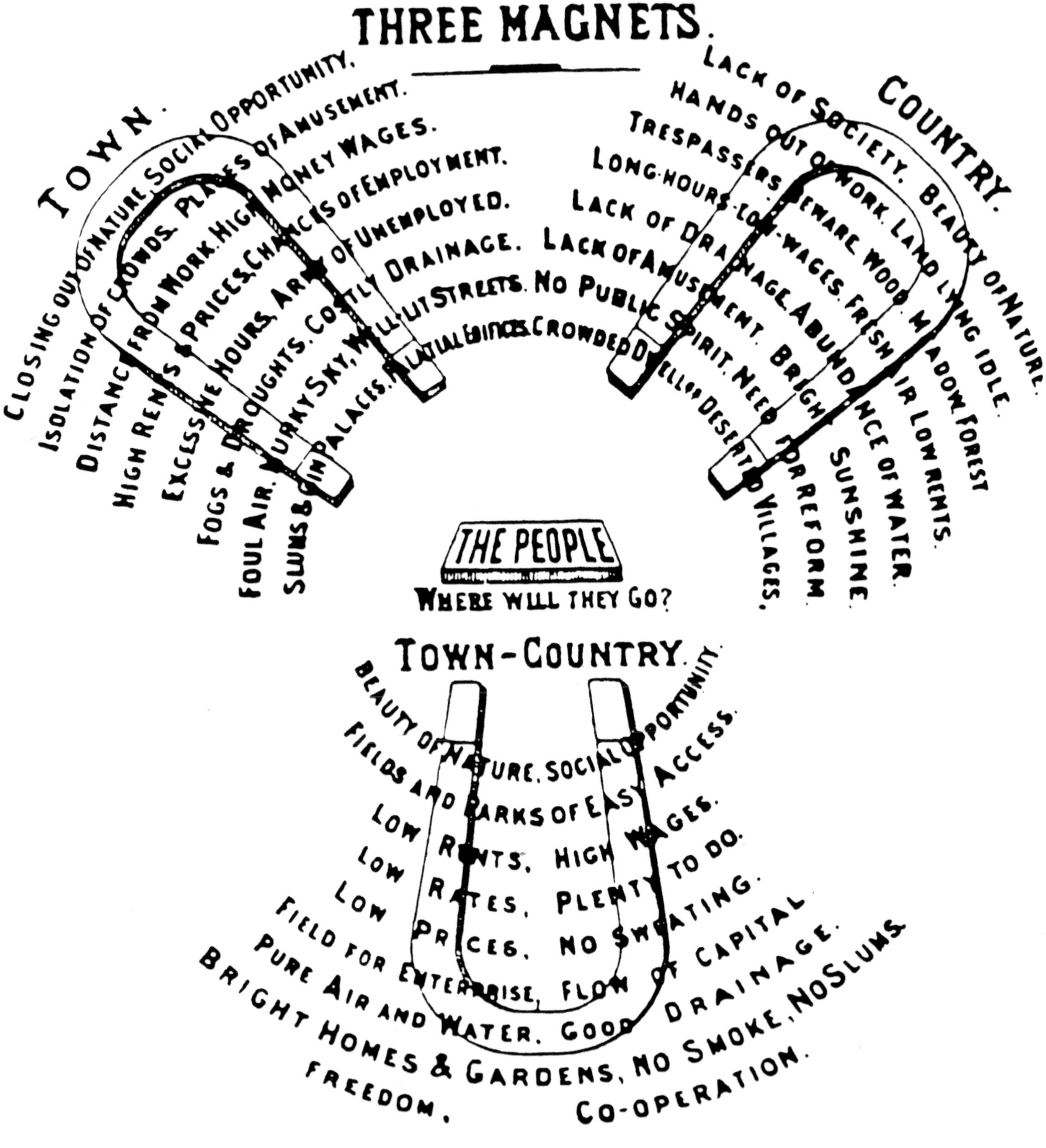

THREE MAGNETS.

TOWN.

CLOSING OUT OF NATURAL SOCIAL OPPORTUNITY. ISOLATION OF CROWDS. PLACES OF AMUSEMENT. DISTANCE FROM WORK. HIGH MONEY WAGES. HIGH RENTS & PRICES. CHANCES OF EMPLOYMENT. EXCESSIVE HOURS. ARMY OF UNEMPLOYED. FOGS & DROUGHTS. COSTLY DRAINAGE. FOUL AIR. MURKY SKY. WELL-LIT STREETS. NO PUBLIC SPIRIT. SLUMS & GIN PALACES. PALATIAL EDIFICES. CROWDED DWELLINGS.

COUNTRY.

LACK OF SOCIETY. BEAUTY OF NATURE. HANDS OUT OF WORK. LAND LYING IDLE. TRESPASSERS BEWARE. WOOD. MEADOW. FOREST. LONG HOURS. LOW WAGES. FRESH AIR. LOW RENTS. LACK OF DRAINAGE. ABUNDANCE OF WATER. LACK OF AMUSEMENT. BRIGHT SUNSHINE. NO PUBLIC SPIRIT. NEED FOR REFORM. CROWDED DWELLINGS. DESERTED VILLAGES.

THE PEOPLE
WHERE WILL THEY GO?

TOWN-COUNTRY.

BEAUTY OF NATURE. SOCIAL OPPORTUNITY. FIELDS AND PARKS OF EASY ACCESS. LOW RENTS. HIGH WAGES. LOW RATES. PLENTY TO DO. LOW PRICES. NO SWEATING. FIELD FOR ENTERPRISE. FLOW OF CAPITAL. PURE AIR AND WATER. GOOD DRAINAGE. BRIGHT HOMES & GARDENS. NO SMOKE. NO SLUMS. FREEDOM. CO-OPERATION.

The Three Magnets
from *Garden Cities of To-Morrow*, 1898
Ebenezer Howard

This is Howard's most famous diagram, which 'proves' the
evils of both city and countryside, in comparison to the
advantages of his hybrid alternative, Town/Country, and
by extension the attraction of his notion of the Garden
City. A damning, if high emotive, view of the city.

Two Tree Island

Nature Reserve

Crow Stone

Chalkwell
Oaze

CENTRAL STATION

Clifftown

Westcliff-on-Sea

Leigh Sand

Sheltered Bay

Wind Surfing

Water

Connection to South End Pier

Maritime/Coastal Housing

Park

Maritime/Coastal Housing

Shopping

Private Beach

Housing

Recreation

Marina

Rural/Wetland Housing

Sport

Floating Houses

New Salt Marsh Nature Reserve

Allhallows-
on-Sea

London Stone

Floating Cafes

Cockleshell
Beach

North Level

Loos
Marshes

Previous pages:

Thames Gateway Vision, 2003

Terry Farrell

As London's population expands, new areas are appropriated for development. None is more tempting than the relatively underpopulated spaces east of the city–the Thames Gateway. The plans for building on this tidal Thames floodplain have ebbed and flowed since major development was first proposed in the late 1980s by Martin Simmons and Peter Hall, but building is frantically underway nonetheless.

One significant vision for the area was initiated by the planner and architect, Terry Farrell, in 2003. Farrell interpreted the entire estuary as a national park with discrete towns of relatively high density scattered in it rather than as continuous development. The plan has had a degree of official endorsement but is far from being the agreed masterplan for the development of the area.

Image courtesy of Farrell & Partners.

Traffic routes had been tackled by Abercrombie, among many others, who had recommended orbital ring roads around London along with two major cross routes. These resulted in some initial schemes, including the north and south circular routes, but it was not until the 1960s that a real appetite for change emerged. During this time, Colin Buchanan was commissioned by the Minister of Transport, Ernest Maples, to work on the report, *Traffic in Towns* (published 1960). In the shortened public edition of the report, Buchanan wrote about "facing a crisis of traffic in towns" and that, even if the population were to remain stable at 1960 levels, "we urgently need motorways inside the cities, not only between them".[21] At the same time, he recognised the importance of a walkable city, although this ultimately led to proposals for segregation of vehicles and pedestrians.

Buchanan's proposals for redrawing the map of London took the form of a study of Fitzrovia in central London, although it is not difficult to imagine the proposals being expanded across the city. One-way motorway-like roads with spectacular interwoven and multi-level junctions would penetrate the city, entangling it with a mass of roads that lead to others, leaving behind small islands of historic fabric. It was never likely to happen, but it did prepare the ground for the later proposals by the Greater London Council (GLC) for an inner London motorway box that, although only partially implemented, has had a very significant impact on the development of the plan of London.

The eventual unpopularity of these big schemes for London, as well as the complex politics of the city and the lack of large areas of land for redevelopment, meant that there was no new official 'plan for London' until after the new millennium. But the desire to both resolve problem and malfunctioning areas of the city, as well as for the grand gesture, led to proposals and substantial developments wherever the imaginations of designers, planners and developers could find a toe hold. Both redundant railway and dock lands became available in the 1980s and plans were drawn up, and in some instances implemented, to create new commercial and even residential areas of London. The area around Liverpool Street and Broadgate stations were substantially built over with a new office city. In contrast, proposals for the King's Cross and St Pancras areas failed to take off and new plans for this area are now only in development.

The main area to be re-planned and re-built were the docks on both sides of the Thames under the auspices of the London Docklands Development Corporation, 1981–1998. Beaux Arts city planning has transformed the West India docks on the Isle of Dogs into a 'Manhattan on Thames' while the Surrey docks in Rotherhithe have become a new residential area. The temptations, to politicians and planners, of the former industrial and relatively undeveloped areas to the east of London remain strong and the move to this area, started by the development of the docks, continues into the optimistically named Thames Gateway.

The decision to elect a London mayor, representing the whole of the Greater London area in 2000, came with the requirement to formulate a new spatial strategy for London—a requirement that generated *The London Plan*, 2004. *The London Plan*'s maps of the city try hard to operate at a strategic and diagrammatic level. Definite lines on maps are avoided so far as is possible—perhaps learning from its predecessors. The London-wide responsibilities of the mayor, and a politic desire for deliberate vagueness, has located the city back into a single plan, despite its multi-layered complexities.

The mayor's first diagram, published in the *Draft London Plan* in 2002, also provides a friendly and inclusive map of the city, with a broad-brush vision thrown in. London needs more such maps if it is to rediscover its identity; maps that explain London to its inhabitants and to others, a London that isn't just a collection of tourist sights or a means of getting from one place to another, but a city that has a collective sense of purpose.

LIVING IN
THE CITY

WORK, SHOPS AND HOME

When the Ordnance Survey was established in 1791, the estate surveyor Thomas Milne, applied for a job with the new organisation. In addition to these credentials, he was described in George Adams *Geometrical and Graphical Essays* as "one of the most able and expert surveyors of the present day".[1] He was however, for whatever reason, turned down. Milne, perhaps in retaliatory response, went on to survey and publish, in his own name, a detailed and radical map of London that pre-empted mapping practice and conventions for decades to come. In making it, he used information from both the Ordnance Survey of the adjacent counties as well as Richard Horwood's map of 1792. Since only one full set of his six sheets is known (in the George III Topographical collection at the British Library) it is possible that he ran into copyright problems with both of these sources—a not unfamiliar issue today—and general publication of the map was suppressed. The techniques he invented and adapted from estate mapping practice had to be rediscovered and reinvented when later land utilisation surveys were undertaken, including Laurence L Dudley Stamp's *First Land Utilisation Survey of Great Britain* in the 1930s.

Milne's breakthrough was the graphical description, through the use of colour-coding, of different land uses and intensities, as well as the age of settlements. This instantly revealed the character of any area and the overall pattern of land use and, as a result, people's lives in London. Authors, from John Stow onward, had described the shape of life in London via literary device, and grappled with spatial ideas such as the change in density, farming practice and lifestyle from the city centre outwards. The Reverend Henry Hunter, in his *History of London and its Environs*, 1811, describes the land around the city as a series of "concentric belts" each with a different agricultural character. On Milne's map this is immediately evident and far more memorably communicated.

Milne was possibly the first to map social data accurately in a way that abstracts it from the purely geographical. His work coincides with that of others, especially William Playfair, 1759–1823—author of *The Commercial and Political Atlas*, 1786, and *The Statistical Breviary*, 1801, and who was credited with the invention of line, bar and pie charts—who were revolutionising the way all types of data could be graphically portrayed.

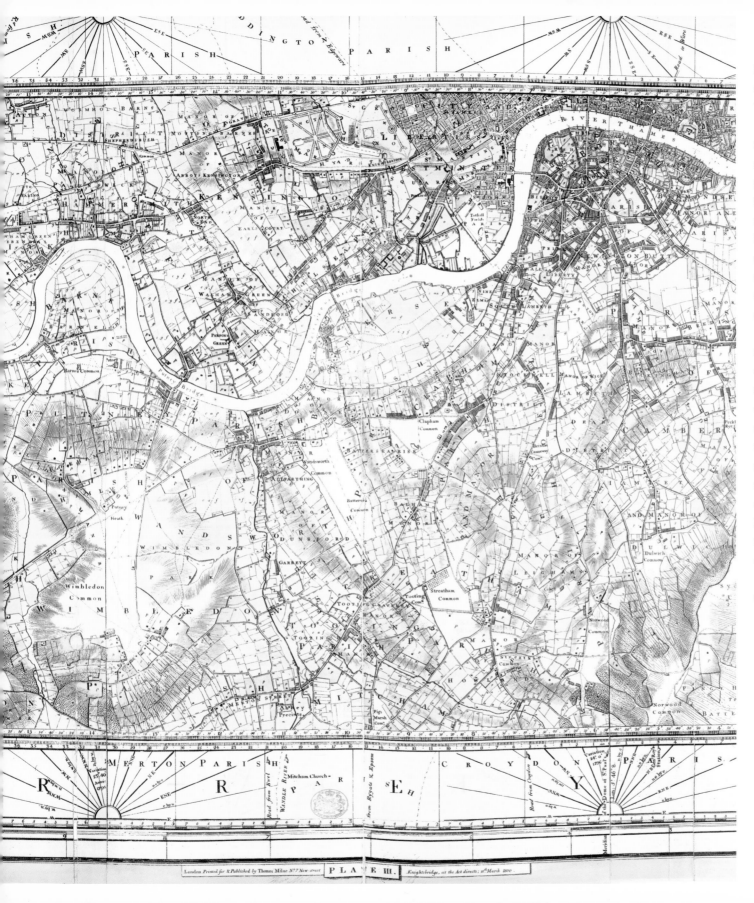

Land Use Map of London
and Environs, 1800
Thomas Milne

Milne's map of London is generated from triangulations
extending from the base line on Hounslow Heath,
precisely measured by General Roy for the Ordnance
Survey and published in 1790, and pre-dates any
maps published by the Ordnance Survey of London.
It therefore can lay a claim to being the first truly
accurate map of London. It may also be why there is
only a single complete copy; as the Ordnance Survey
may have stepped in to ensure that Milne was never
able to publish the map.

Field boundaries are meticulously plotted and
then annotated with both a letter key and coloured
wash to show the land use and geology of each,
such as arable, meadows, market gardens, marshes,
woods and parks. Milne's accuracy and attention to
detail makes it possible to read from the map enough
information to give a rounded picture of how life
might be lived in an area of the city; where supplies
of food and building materials were obtained and
areas of work, as well as those of leisure and homes.
In this map, Milne established a series of standard
tools for mapping the city and its hinterland that
would continue in use until the present day.

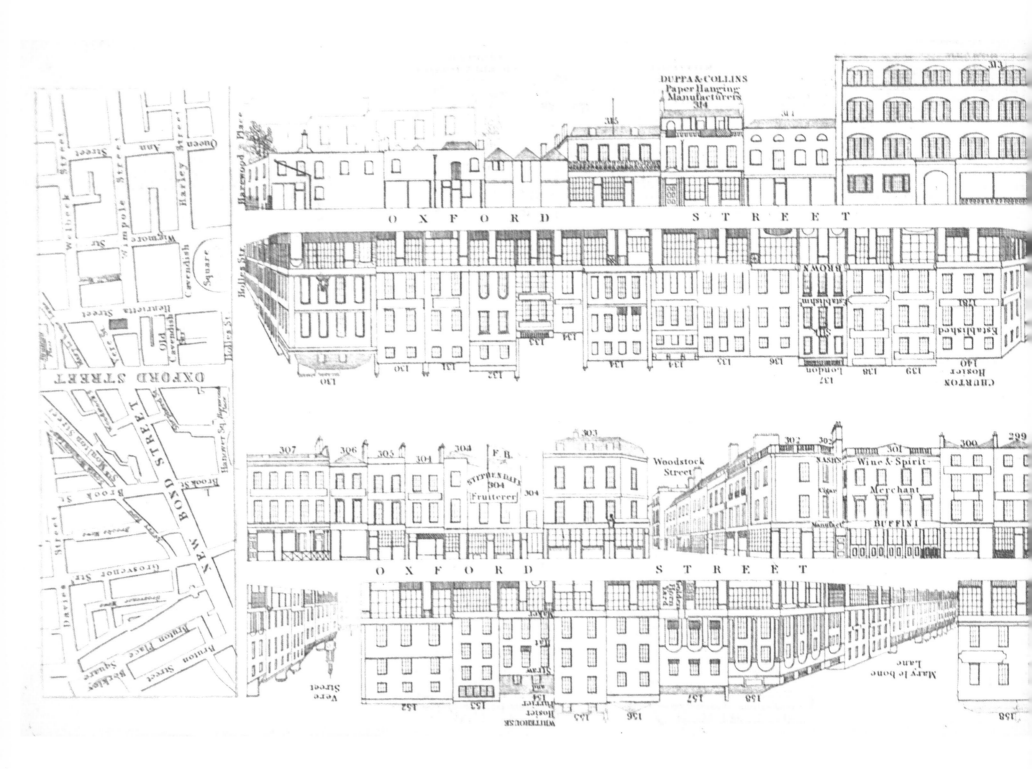

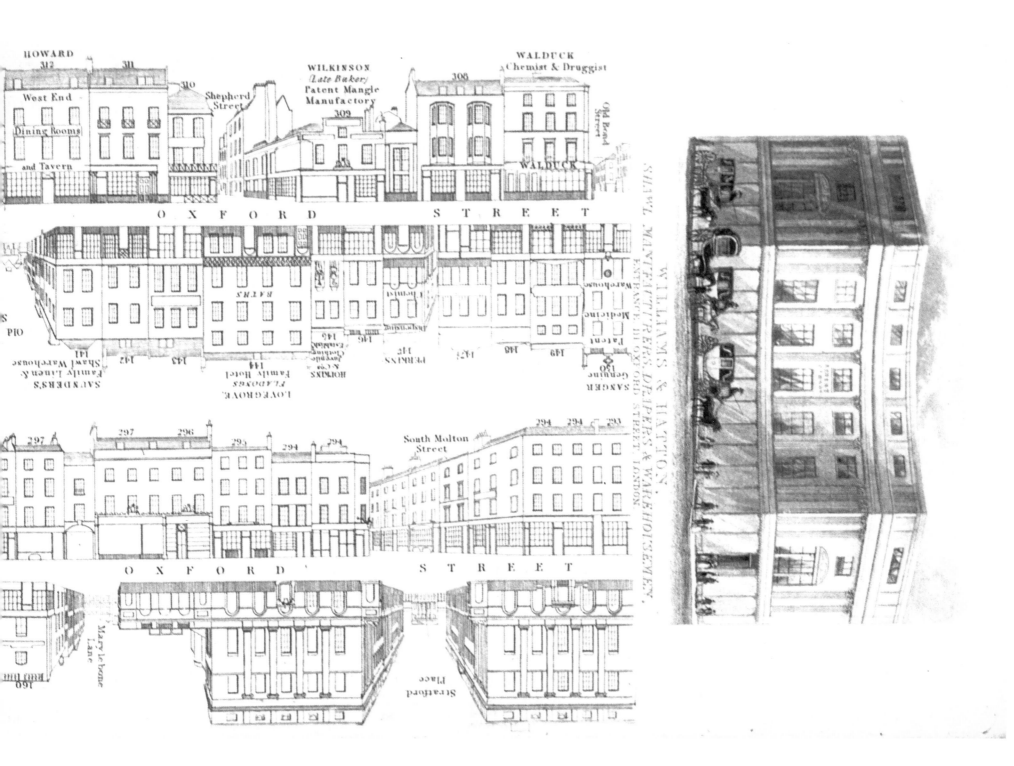

Previous pages:

London Street Views, part 34;
showing Oxford Street, 1838
John Tallis

Street views had long been a staple product of
printmakers and sellers, but Tallis' project was far
more ambitious. He chose to map each street showing
the building elevations with perspective views down
cross streets. Tallis mapped over 175 London streets,
including this one of Oxford Street, in this manner.

His street views provide detailed information
making it possible to recreate the way each street was
used by the people who lived and worked in individual
buildings—houses, shops, manufactories, offices etc.
Tallis' concept of street mapping is still being revisited
today, with contemporary street frontages available
online—or in retail street directories—produced to aid
both shoppers and armchair explorers.

Image courtesy of the Guildhall Library, City of London.

Coal Exchange Invitation, 1849

London has always been a city of both work and
pleasure and this map, a ticket for the opening ceremony
of the New Coal Exchange on 30 October 1849, printed
on a folded paper doiley, gives an unusual insight into
the business side of the city, even if it is promoting a
celebration. The map shows the site of the Exchange,
sending out a clear message about its location at the
heart of the greatest commercial city in the world.

Image courtesy of Science & Society Picture Library.

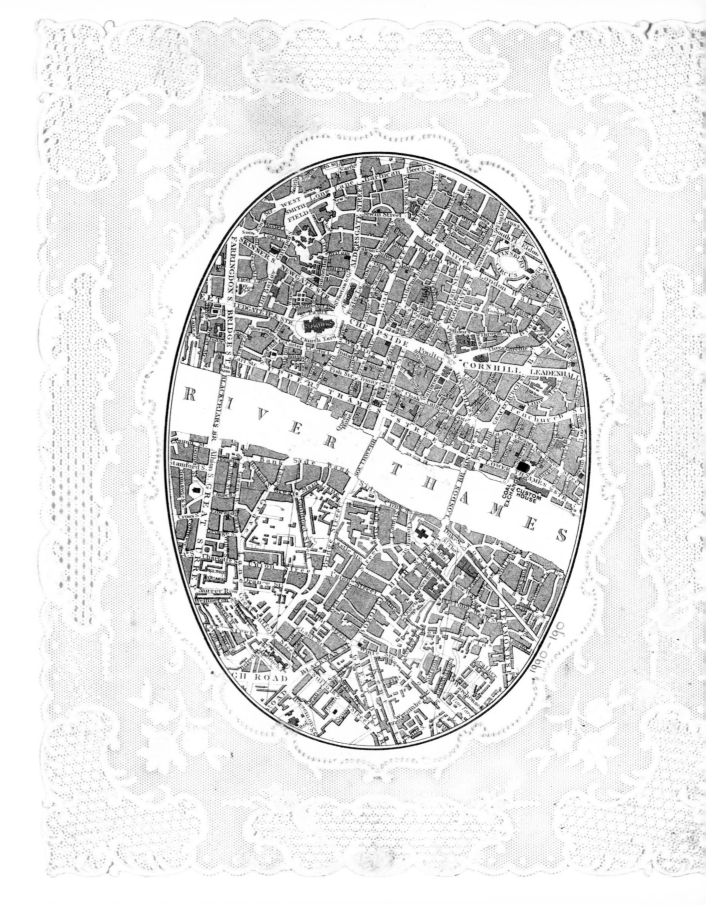

From this moment on, there could be no looking back and maps would eventually become essential tools for both the analysis and explanation of concepts and ideas. As Edward Tufte notes: "Often the most effective way to describe, explore and summarise a set of numbers—even a very large set—is to look at pictures of those numbers."[2] But despite these early beginnings, London, and the world at large, would have to wait. It was almost 90 years until Milne's data-mapping innovations became commonplace techniques.

During the nineteenth century, both merchants and shoppers were assisted by maps showing the location of shops and places of entertainment. The locations of pubs were mapped for entirely different reasons by the National Temperance League, hoping to shock politicians by their very prevalence (and presumably popularity) into shutting them down. Maps also become ubiquitous enough to be used as invitations for single events but they were little used for serious social analysis until a new spirit of enquiry appeared—one perhaps characterised by the first appearance of Sherlock Holmes

in *Beeton's Christmas Annual* in 1887. Snow's plans, in 1854 and 1855 retrospectively, of mapping cholera statistics can be seen as an aberration in the period and it is perhaps not surprising that he found it difficult to find anyone that would take his ideas seriously.

In contrast, Holmes' faith in deductive reasoning based on evidence and observation, must have struck a chord with Charles Booth whose *Descriptive Map of London Poverty* began to be published in 1889, and were followed by a further 14 years of investigation and detailed map-based publications. Booth's work ushered in a new period of enquiry and an explosion in academic studies, although it is only towards the end of the twentieth century, with the aid of computer mapping techniques and the Internet, that the information he and his associates gathered has become easily accessible to the public again. The *First Land Utilisation Survey of Great Britain* of the 1930s, coordinated by Laurence Dudley Stamp and undertaken on the ground by a quarter of a million volunteers, including students and schoolchildren overseen by geography teachers, was intended

for academic purposes but turned out to be vital resource for food production and planning policy both during, and following, the Second World War.

The follow up—*the Second Land Utilisation Survey of Britain*—was carried out in the 1960s by Alice Coleman and provided essential information, along with data from the census and other sources, for the *Atlas of London and the London Region*. The *Atlas* contains over 60 maps of London illustrating a wide range of statistics including employment, population, residential tenures, industrial and economic activities as well as fertility and levels of overcrowding. The *Atlas* pioneered many new representational techniques that make the data highly accessible, but its unwieldy size and general awkwardness prevent meaningful access to more than a few researchers. Only when *A Social Atlas of London* was published did such information become widely disseminated.[3]

Computer databases linked to maps can now produce, make available and keep updated such information almost automatically.

Examples include a house price map produced by myhouseprice. com—a map that provides far more information on the state of London than simply a report on the price rates of property. We can expect to see many more such mapping exercises in the future, which take forward the transformation in the use of maps begun by Thomas Milne at the end of the eighteenth century, into our information age.

London House Price Map
myhouseprice.com

This map, computer generated from thousands of individual house sales, can assist any Londoner, or expert valuer, to make an educated guess at the price of their property. For the sociological observer it also provides a vivid portrait of the levels of wealth, as well as relative poverty, existing in close proximity. It is a modern day version of Booth's poverty map, although the interest in this instance is in just how much householders are 'notionally' worth.

Image courtesy of myhouseprice.com.

Plan of the Crystal Palace and grounds, Sydenham, 1911

Joseph Paxton's 'Crystal Palace', the glazed exhibition hall originally constructed for the Great Exhibition in Hyde Park, moved in 1854 to Upper Norwood in South London and became the centrepiece of a new park. Its nickname, affectionately bestowed on it by *Punch* magazine, became the official name of the whole area.

1911 was the year of the Festival of Empire, held to mark the coronation of George V. Despite being used as the location for such an event, the building suffered lack of funds for maintenance shortly thereafter, and was put up for auction. This plan was published by the auctioneers Knight, Frank & Rutley to mark its sale.

Image courtesy of Science & Society Picture Library.

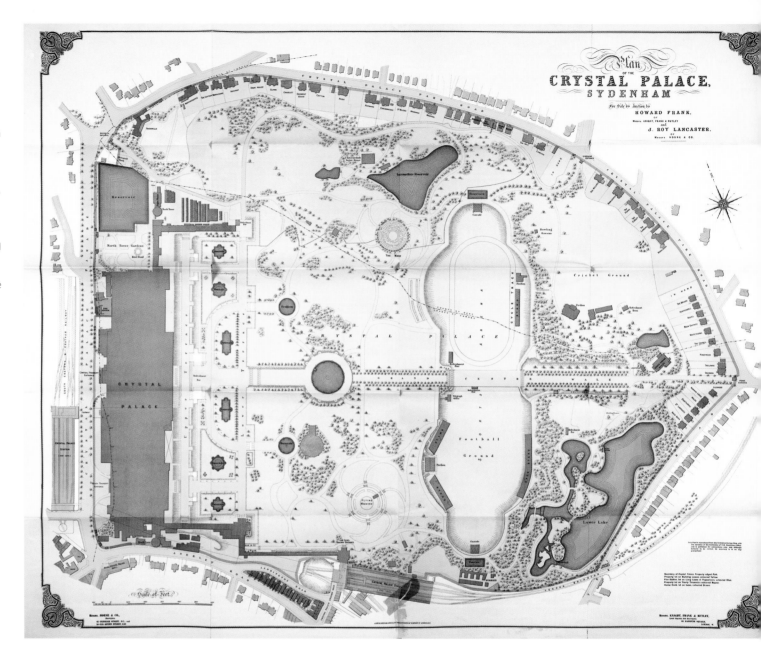

Events and Entertainment

There they are, at the bottom of William Smith's 1588 panorama of Elizabethan London; the theatre and the bear and bull-baiting arenas. Not within the boundaries of the city, where such activities were forbidden, but in louche Southwark amongst the taverns and brothels of Bankside. Other theatres existed in Shoreditch (The Curtain) and even closer to the city centre; within former monastic grounds that escaped the control of the city, such as the Blackfriars Playhouse and the Holywell Priory. The standard impression we now have of Elizabethan London; that created by Shakespeare, Marlow and Jonson, inevitably has a prominent theatre—the Globe—situated in the foreground, just as it is in Smith's image.

Entertainment in London is a part of its way of life, its essential character and always inextricably connected to its celebrated seedy side. But by 1746, the city's theatres are barely visible on John Rocque's map. Charles II had granted warrants to only two theatre companies, whose premises off Bow Street and in Lincoln's Inn Fields are discreetly labelled. Far more obvious are the Pleasure Gardens of Ranelagh, Vauxhall Spring Gardens and Marylebone that became the most popular form of mass entertainment during the period, for the respectable and dissolute alike.

Perhaps because many of London's mapmakers were, like John Rocque, trained as landscape and estate surveyors, the gardens and with them, London's many parks, have been extensively surveyed and mapped. One such example includes a very early, 1400 chart granting ownership of Faux Hall to Roger Damory and his wife Elizabeth, known through a much later copy in the archives of Canterbury Cathedral. Vauxhall Gardens, having featured so large in London's life—in reality as well as in its fiction—was finally closed when its owners went bankrupt in 1840. It was broken up and sold in 1842.

By the mid-nineteenth century, when trains started bringing vast numbers to central London on excursions, the smaller scale pleasures of places like Vauxhall probably seemed hopelessly outdated. Much larger events were called for: events that reflected the Imperial standing of Victorian Britain, such as the Great Exhibition of 1851, which became the model for all those that followed. While Paxton's great greenhouse structure for the Great Exhibition had become too cherished to destroy and was re-erected in an enlarged form in 1854 on Sydenham Hill (in the area re-christened Crystal Palace), subsequent exhibitions were held in South Kensington in 1862, 1871 and 1874, finally leading to the great museum district of Albertopolis. The Victorians discovered that they liked large

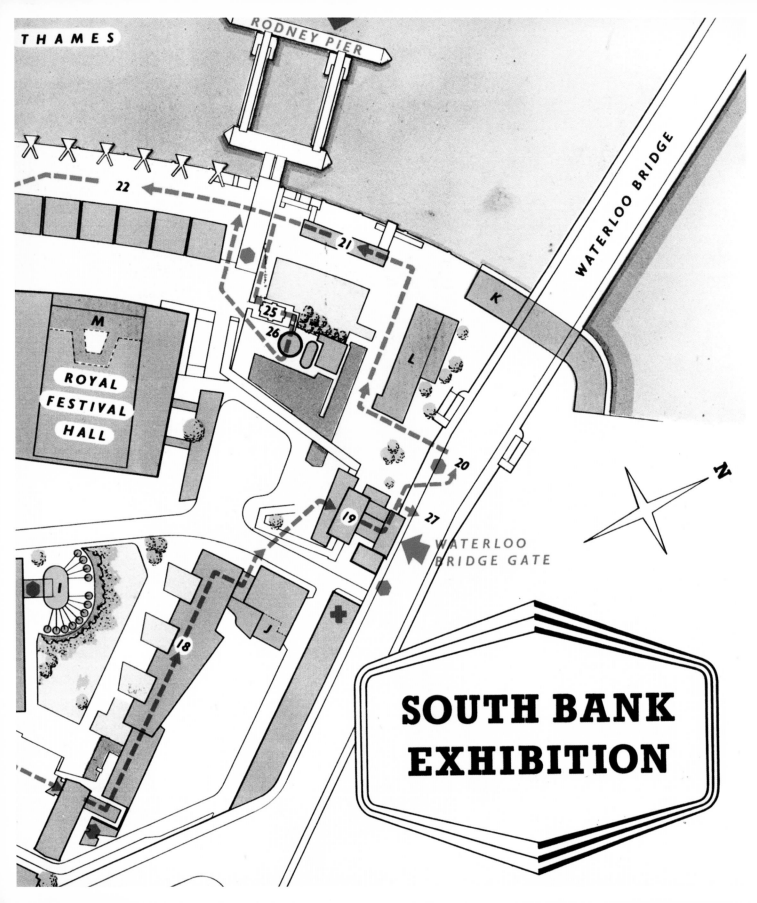

THAMES

RODNEY PIER

22

21

25

26

M

ROYAL
FESTIVAL
HALL

20

19

27

WATERLOO
BRIDGE GATE

I

J

18

K

L

WATERLOO BRIDGE

N

SOUTH BANK
EXHIBITION

Left and opposite:

South Bank Exhibition

Festival of Britain, 1951

Ian Cox

The Festival of Britain continues a tradition of populist events stretching from the Great Exhibition of 1851, through to the Millennium Experience of 2000. Like the British Empire Exhibition in 1924, it had an explicit role in the recovery of the country after an exhausting and draining war. The many maps associated with the Festival explicitly support this aim by encouraging people to visit and travel around London.

These images from the official catalogue to the Festival of Britain show how maps had become a commonplace way of describing how to navigate around such an event. All the main pavilions have individual plans as well as the main plan of the entire exhibition.

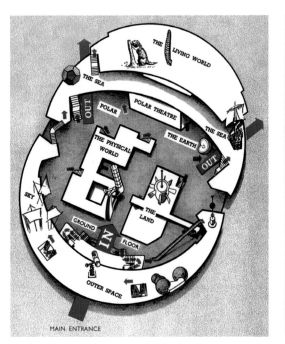

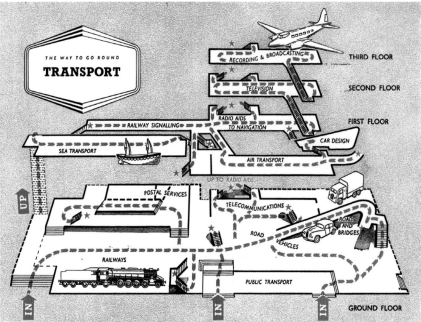

communal gatherings and, despite their worst fears, that the vast crowds were generally good-natured and well behaved. It went without saying that such events should also be educational, 'improving' and above all, good for the moral fibre of the greatest country in the world.

Similar moral values and concern about the behaviour of the proletarian masses, were behind the simultaneous laying out of new parks which would, in the words of the 1833 Select Committee on Public Walks, "assist to wean them from low and debasing pleasures".[4] Following the success of Regent's Park at the beginning of the nineteenth century—part of John Nash's great urban vision that linked it to St James' Park and the heart of royal London, and which included the opening in 1828 of the Zoological Society, parks had become all the rage. The problem was to find and afford the land to create them, as the metropolis spread rapidly outwards, and the value of land suddenly upward. Victoria Park in Bow, designed by James Pennethorne (Nash's adopted stepson), was opened in 1845 and became, according to Charles Booth: "The arena for every kind of religious, political or social discussion."[5] Battersea Park in South London, also by Pennethorne, opened in 1858 replacing a wild and anarchic Sunday fair, and other morally suspect activities, with far more

proper and respectable sports and recreational pursuits. After 1856, and the creation of the Metropolitan Board of Works (MBW), Southwark Park was added to the list. It was also the last park to be created in the centre of London until the development of Mile End Park in the 1990s.

All of these improvements were accompanied by numerous maps and plans, during their design and development. These were often produced under intense public scrutiny but later functioned as useful information once the attractions were open to visitors. The anticipated enjoyment and festive spirit of 'a day out' at an exhibition, or in the park, spills over into the design of the maps, inspiring a graphic freedom missing from many equivalent contemporary maps. This is as true for the great twentieth century events, such as the British Empire Exhibition of 1924 and the Festival of Britain in 1951, as for those of the Georgians and Victorians before them. It is only the lacklustre presentation of the equivalent Millennium Experience in 2000 that breaks the pattern, although the London Olympic Games of 2012 look set to return to form with the early plans for the Lower Lea Valley, and the eventual creation of another substantial park in London, whose ambitions could match those of Nash—if only the promoters hold their nerve.

Site Map for the British Empire Exhibition, 1924

Following the success of the Great Exhibition in Hyde Park, in 1851, London played host to many more showpiece events designed to celebrate the accomplishments of the British Empire. The British Empire Exhibition hosted exhibits from 56 of the 58 countries it governed and was intended, in part, to rebuild the spirit of the Victorian period in the aftermath of the First World War. The plan, however, is not particularly synonymous with a great Imperial setpiece, especially with its use of circus motifs and the inclusion of the old Wembley Stadium (with its connotations of Roman games and contests).

Image courtesy of the Museum of London.

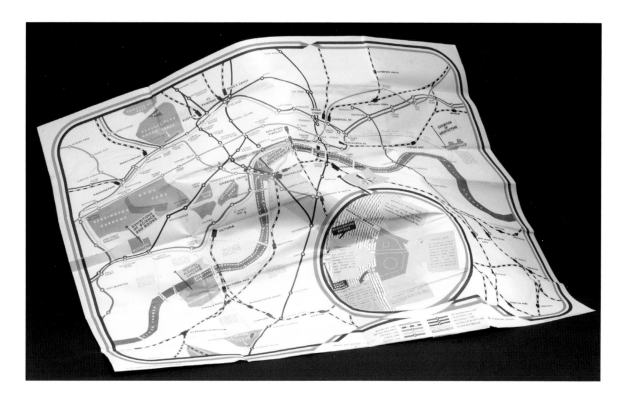

Welcome to London, Festival of Britain, British Railways and London Transport Pocket Map, 1950

The Festival of Britain in the Waterloo area of London's South Bank was conceived as an uplifting popular event in the wake of the Second World War. Its prime motivator, the Deputy Leader of the Labour Party, Herbert Morrison, described the Festival as "a tonic to the nation".

This printed map, issued by London Transport and British Railways, was intended to show the public how to reach the festival site by public transport, but also highlights other areas of the Festival, at South Kensington (science), Poplar (housing) and Battersea (sculpture). The map neither follows the traditions of the iconic London Transport maps nor the heightened modernism championed by the Festival itself.

Image courtesy of Science & Society Picture Library.

London Marathon Course Map, 2007

The London Marathon has become an essentially urban experience, deliberately passing well-known landmarks and using them to animate its route. As a result, the map of the course is one of its critical components, even if it is more suited to spectators than the competitors themselves.

Image courtesy of London Marathon Limited.

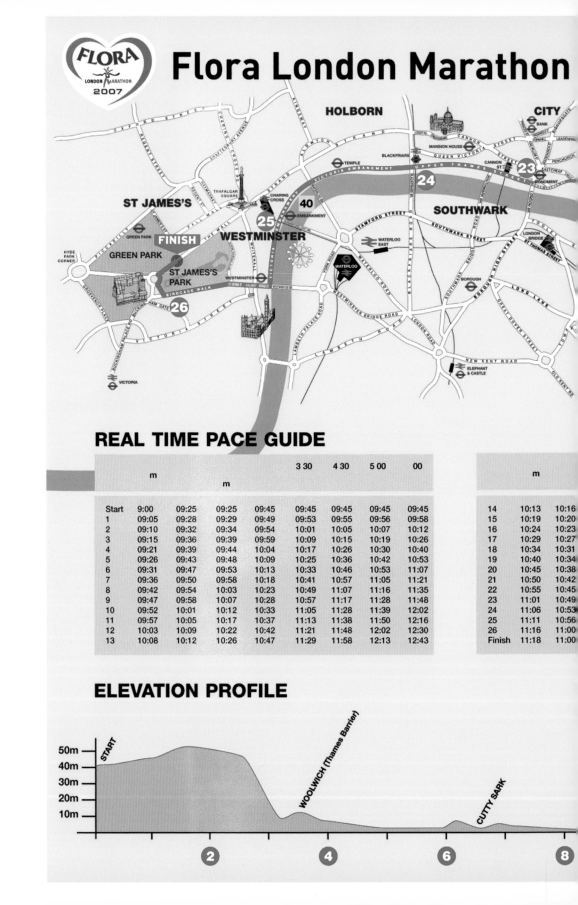

007 Course Map

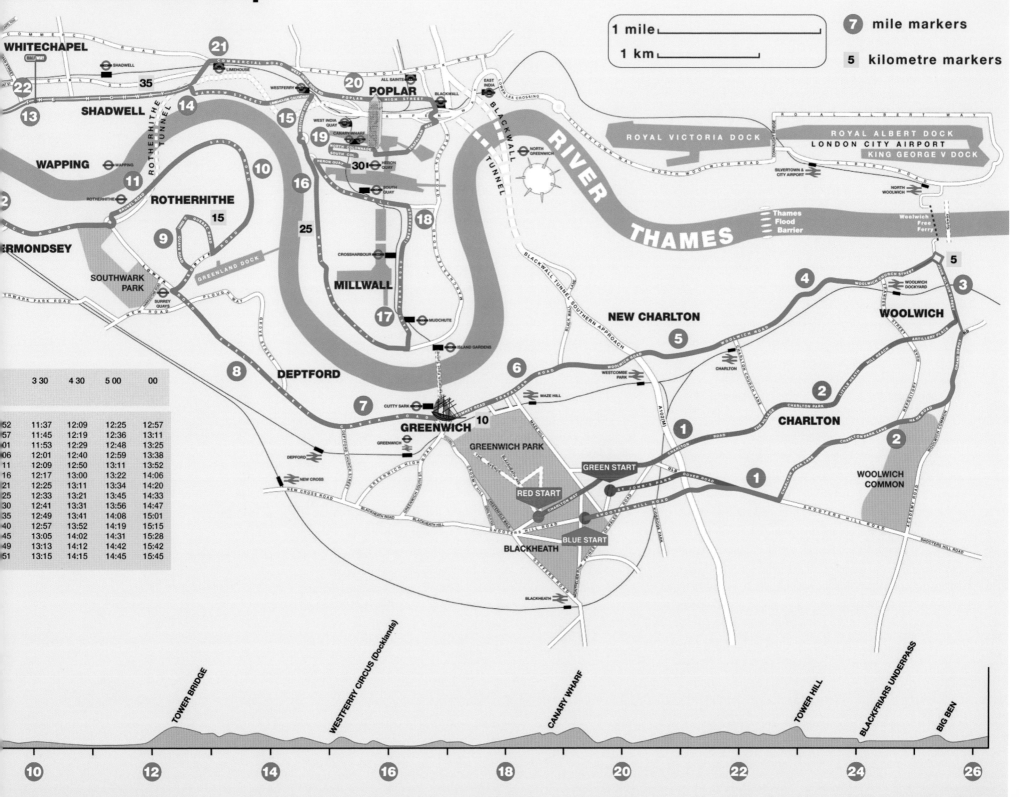

| 1 mile | |
| 1 km | |

7 mile markers
5 kilometre markers

WHITECHAPEL · SHADWELL · WAPPING · BERMONDSEY · ROTHERHITHE · DEPTFORD · GREENWICH · POPLAR · MILLWALL · NEW CHARLTON · WOOLWICH · CHARLTON · BLACKHEATH

ROYAL VICTORIA DOCK · ROYAL ALBERT DOCK · LONDON CITY AIRPORT · KING GEORGE V DOCK

RIVER THAMES

Thames Flood Barrier · Woolwich Free Ferry

GREEN START · RED START · BLUE START

GREENWICH PARK · SOUTHWARK PARK · WOOLWICH COMMON

	3 30	4 30	5 00	00
52	11:37	12:09	12:25	12:57
57	11:45	12:19	12:36	13:11
01	11:53	12:29	12:48	13:25
06	12:01	12:40	12:59	13:38
11	12:09	12:50	13:11	13:52
16	12:17	13:00	13:22	14:06
21	12:25	13:11	13:34	14:20
25	12:33	13:21	13:45	14:33
30	12:41	13:31	13:56	14:47
35	12:49	13:41	14:08	15:01
40	12:57	13:52	14:19	15:15
45	13:05	14:02	14:31	15:28
49	13:13	14:12	14:42	15:42
51	13:15	14:15	14:45	15:45

TOWER BRIDGE · WESTFERRY CIRCUS (Docklands) · CANARY WHARF · TOWER HILL · BLACKFRIARS UNDERPASS · BIG BEN

10 · 12 · 14 · 16 · 18 · 20 · 22 · 24 · 26

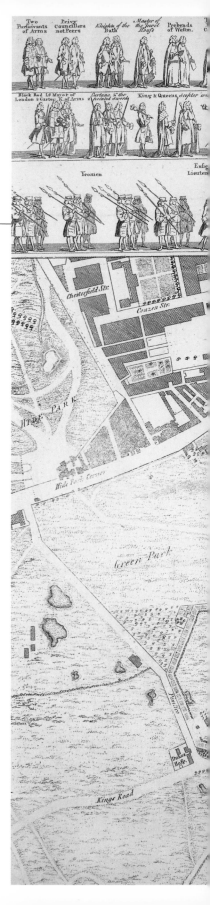

PROCESSIONS AND PROTEST

London in 1820, furiously growing and expanding, was also in trouble: "Confidence in London hit a low: its problems of growth—crime, destitution, epidemic disease, overcrowding—seemed to be on the verge of overwhelming the city."[6] Crime and disorder were blamed for what were essentially problems of extreme population growth. In 1919, in the wake of the Peterloo massacre in Manchester, the government brought in the notorious Six Acts, to outlaw sedition and action for reform. In the words of the preamble to the Acts: "Every meeting for radical reform is an overt act of treasonable conspiracy against the King and his government."[7]

London fears a mob; but since the Gordon Riots of 1780, most threats of violence on the streets had turned out to be simply threats, or exaggerated fear. However, in 1820, insurrection was in the air. A conspiracy to murder the whole cabinet and the new King, following the death of George III in January, was exposed in February in Cato Street (just off the Edgware Road) and the plotters were later either hanged or transported. Years later, the plot was revealed to have been fomented by a government agent provocateur, George Edwards, but the hysteria was real enough.

The early 1820s was also a period of international turmoil with, from 1820–1822, revolutions in Spain, Italy and Greece and the wholesale liberation of South America. In 1820 Metternich,

the Austrian statesman and diplomat, wrote to Tsar Alexander: "Governments, having lost their balance, are frightened, intimidated and thrown into confusion."[8]

To add to the disorder, George III died on 29 January 1820. Despite the febrile atmosphere, the chief concern of the new King, George IV, (the extravagant and obese former Prince Regent) was to dissolve the marriage to his estranged and exiled wife, Caroline of Brunswick, before the coronation, to prevent her ever becoming Queen. Popular opinion in London backed Caroline's cause against that of the new King, and radicals, led by Matthew Wood, MP for the City of London, persuaded her to return to England to be present at the coronation. Caroline was eventually and dramatically turned away at the doors of Westminster Abbey by prizefighters hired for the occasion. She fell ill later that night and died three weeks later.

London, having acclaimed Caroline on her return from exile with some of the largest demonstrations ever seen in the city, returned to the fray for her funeral procession in August 1821. During the chaos and anarchy of the procession, troops—in response to being pelted with stones—shot dead two demonstrators. The Prime Minister of the time, Lord Liverpool, was so panicked by events he wrote to the French writer and diplomat, Chateaubriand, stating: "One serious insurrection in London and all is lost."[9]

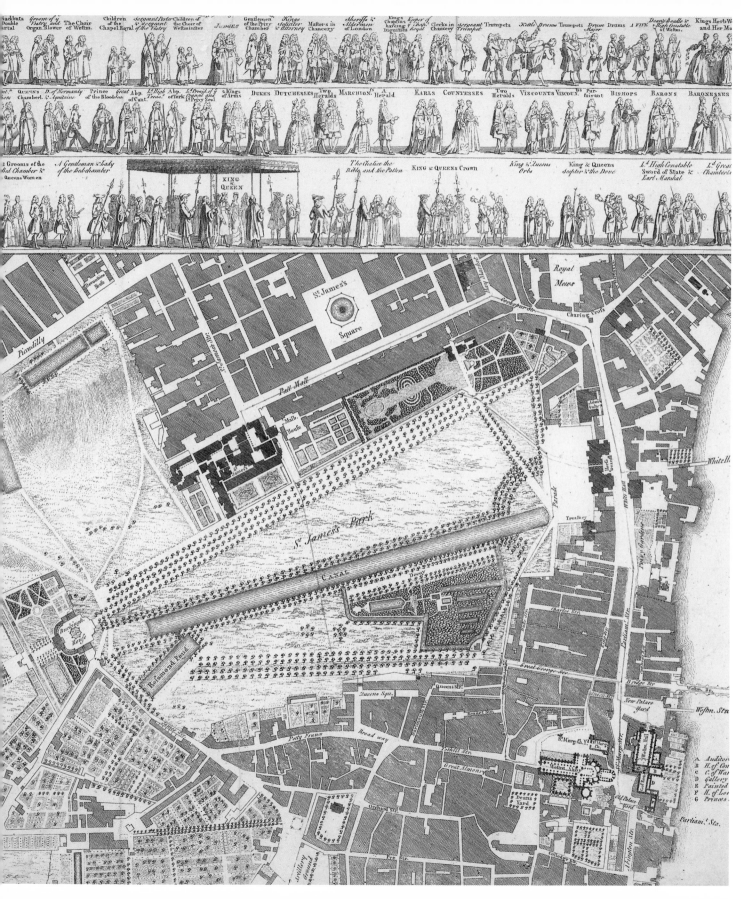

Procession for the Coronation of George III, 1761

The Coronation of George III was an expansive affair, and the plan of the procession shows its formation as a figure of eight through St James' Park, along The Mall and back to Westminster Abbey across the top of the plan is a display of 'the great and the good' (in order of social precedence), who accompanied the new King on his way to the ceremony. The title boasts that this was the 'usual' course of events during a Coronation when, in truth, each ceremony reflected the particular mood of London at the time.

Right and opposite:

Plans of Westminster showing the arrangements for the Coronation of George IV, 1821
James Wyld

The coronation of George IV enabled London radicals to use the cause of his exiled wife, Caroline, as the focus for renewed anti-government protests. Unlike the processional route of James II, George IV's Coronation wasn't lined by troops but heavily flanked by grandstands for the invited audience. Caroline managed to penetrate the event as far as the Abbey doors, where she was forcefully turned away. Her death shortly after resulted in further, and far more violent, riots. This was a turning point in relations between the aristocracy and the masses, who henceforth knew the extent of their power.

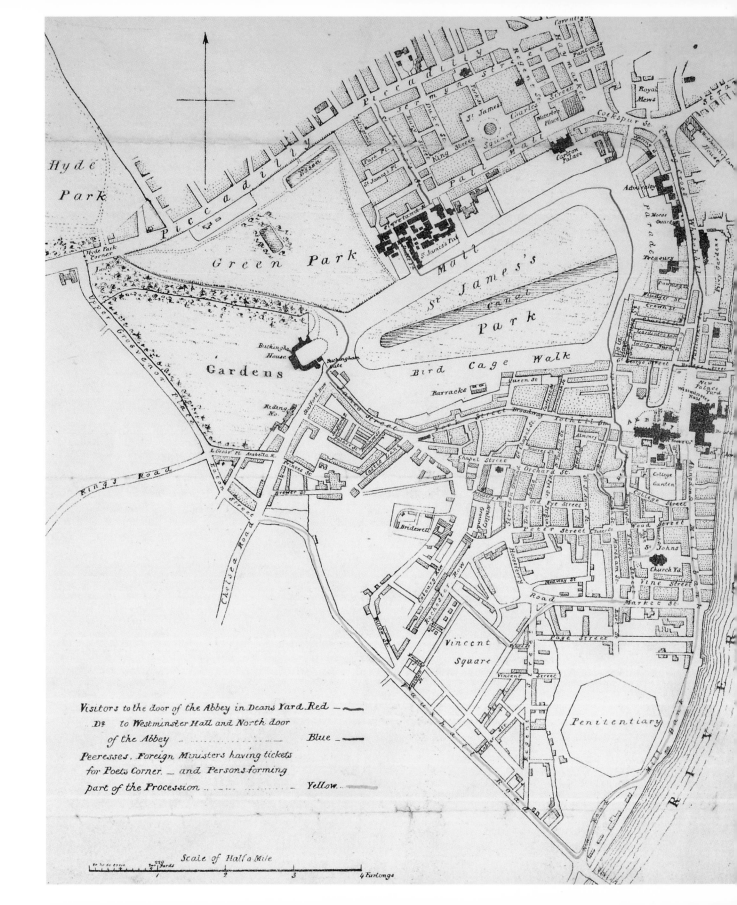

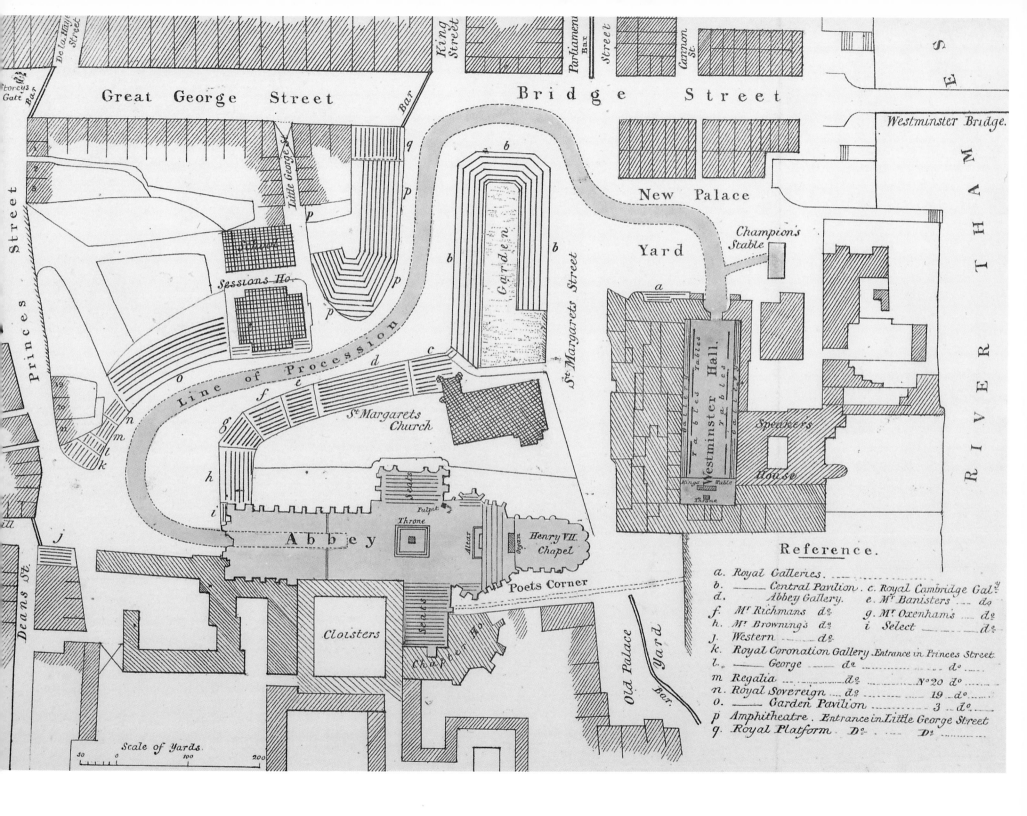

Great George Street

Bridge Street

Westminster Bridge.

RIVER THAMES

New Palace Yard

Champion's Stable

Westminster Hall.

Tables Tables Gallery Tables Tables Gallery

Kings Table Table

Throne

Speakers House

Reference.

a. Royal Galleries.

b.   Central Pavilion   c. Royal Cambridge Gal.

d.   Abbey Gallery.   e. Mr Banisters   do

f. Mr Richmans do.   g. Mr Oxenhams do.

h. Mr Browning's do.   i Select   do.

J. Western   do.

k. Royal Coronation Gallery Entrance in Princes Street.

l.   George   do.   do.

m Regalia   do.   No 20 do.

n. Royal Sovereign do.   19 do.

o.   Garden Pavilion   3 do.

p. Amphitheatre. Entrance in Little George Street.

q. Royal Platform Do.   Do.

Garden

St Margarets Street

King Street

Parliament Street

Cannon St.

Princes Street

Little George St

Sessions Ho.

Line of Procession

St Margarets Church

Abbey

Throne

Altar

Organ

Pulpit

Seats

Seats

Seats

Henry VII. Chapel

Poets Corner

Cloisters

Chapter Ho.

Old Palace Yard

Deans St.

Scale of Yards.
50   0   100   200

The Caroline Riots, named after an unusual heroine for an anti-royalist protest, had both radicals and reformers fighting for her cause and combined the crowd-pulling potential of both a royal procession and a political demonstration.

Due to their ability to attract large numbers of people onto the streets, both processions and protests were regular subjects for mapmakers and annotators. James Wyld's plan for George IV's coronation route reveals none of the eventual tumult or indeed the pantomime pageantry of the occasion, concentrating rather on the procession from Westminster Hall to the West door of the Abbey—although it does reveal the extraordinary level of organisation put in place to keep the various different classes of attendees apart.

London has maintained its role as a venue for protest and pomp ever since; with maps inevitably accompanying each event. Whether in anger at the defeat of the Reform Bill in 1832, in support of Chartism in the 1840s or the Suffragette Movement, at the beginning of the twentieth century, or against the wars in Vietnam, 1966, or Iraq, 2003, protests have continued to fill the streets of London. Equally, London has hosted frequent state occasions with similarly large numbers coming to cheer or mourn, whether at the Queen's Coronation in 1953 or the funeral of Princess Diana—the contemporary Caroline of Brunswick—in 1997.

However, since 1820, the traditions of pageantry and protest have kept a formal distance from one another.

The maps accompanying such events reveal their narrative power; an ability to tell a story over and above raw information about the streets and the lives lived on them. They have a temporal quality that encourages a series of events to be described and imagined. We are familiar with such maps being used in novels and films to show the progress of characters across the globe. Such maps are rarer in describing the 'real life' of cities, and the early maps of procession and protest are the precursors of those imaginative endpapers in adventure stories, or Hollywood's shorthand technique for transferring leading characters from continent to continent.

Today, newspapers are far more likely to produce maps of royal events than to stir up insurrection on the streets, but one thing hasn't changed: the government's fear of mass movements and attempts to organise sedition. In 2005, amid fears for the safety of government, unauthorised gatherings, marches or demonstrations were banned from within an area one mile around the Houses of Parliament. Accompanying the legislation is a map to show where such spontaneous activities are outlawed. Gratifyingly, protest against such draconian measures is finding ways to make maps tell their own story in response.

TO HYDE PARK!

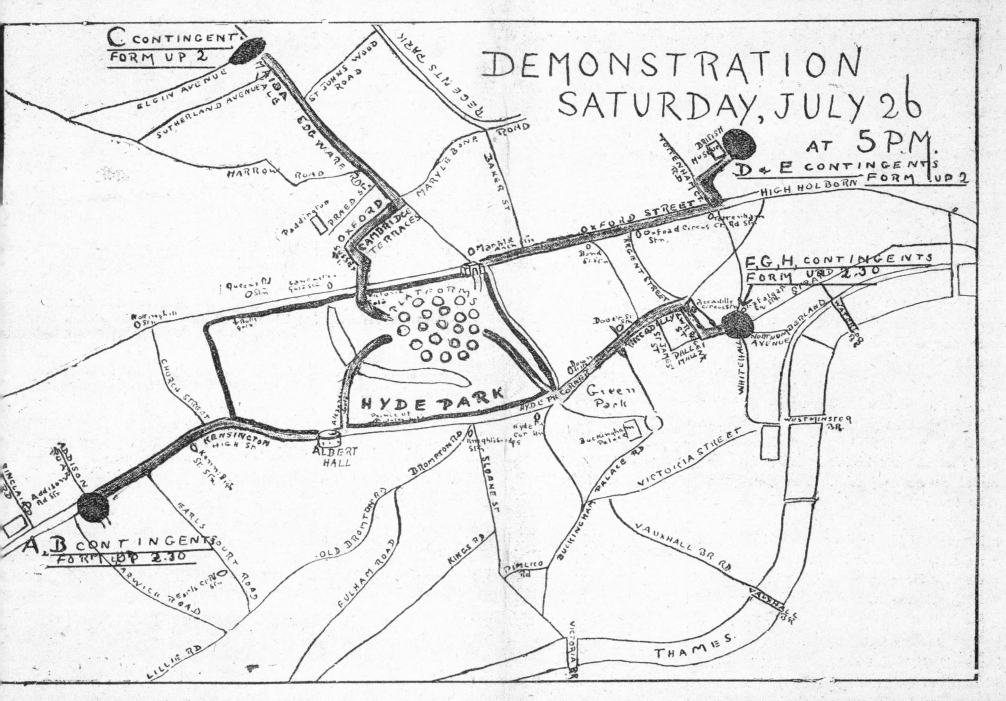

Map of Hyde Park and environs, showing the points of assembly and routes of the four great processions joining in the monster demonstration, at 5 p.m., in the Park to-morrow (Saturday), which is the culminating feature of the National Pilgrimage through England, organised by the National Union of Women's Suffrage Societies.

POLICE AND CRIME

London had long been proud of not having a police force. In 1822, a committee chaired by the then Home Secretary, Sir Robert Peel, declared that a continental-style force would be "odious and repulsive… it would be a plan which would make every servant of every house a spy on the action of his master, and all society spies on each other".[10] In 1750, the author and magistrate, Henry Fielding, initiated a band of 'thief-takers', the first example of such a force in London. The group subsequently became known as the Bow Street Runners and, along with constables from other similar police offices, rose to number approximately 450, reporting directly to Peel.

In 1828, he commissioned a select committee to report on the state of crime in London and Middlesex (already knowing the outcome). In 1829, having regaled Parliament with tales of the criminal underclass of London, Peel had the Police Reform Bill passed, and the Metropolitan Police Force was established with a jurisdiction across the Greater London area. The area under its control was initially within a seven mile radius of Charing Cross. In 1839, this was extended to 15 miles, encompassing an area of approximately 700 square miles and a force of 4,300 officers. The Metropolitan Police jurisdiction, although not including the City of London, was the first to recognise the reality of the effective size of the wider city.

Despite a newly active Police Force, nineteenth century London was far more famous for its crime than for the ability to prevent it. The crime solvers, with which we are most familiar from this period, are more likely the fictional creations of Charles Dickens and Wilkie Collins, as well as the later Arthur Conan-Doyle, whose characters Sherlock Holmes and Dr Watson battled city crime for over 30 years. In Victorian London, the public were indeed terrified of crime, but by 1838 crime rates in the city had fallen dramatically and public confidence in the new system of policing appeared strong. Concern could then pass on to issues of vice and morality, although the public continued to be fearful of violent street robbery, or garrotting, in the 1850s and 60s, and armed burglary in the 1880s. Campaigning journalists and writers, including Henry Mayhew and William Thomas Stead, successfully ensured that issues relating to crime—child prostitution for example—and the criminal underclass were of continual concern to the general public. Those that read London newspapers and magazines were fed a constant diet of crime stories and warnings as to the dangers of the city. In the fourth book of Mayhew's *London Labour and the London Poor*, one of his researchers, Bracebridge Hemyng, warns: "London crime has arrived at a frightful magnitude; nay, it is asserted that nowhere does it exist to such an extent as in this highly favoured city."[11] In truth, "in an average year in

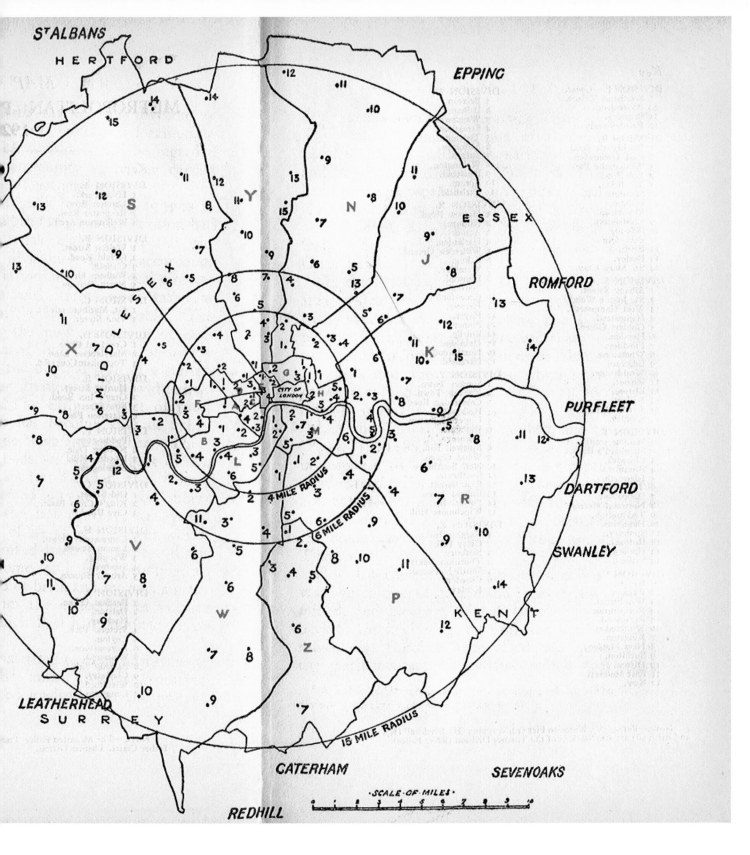

Map of the Police Divisions of London, 1929
Scotland Yard

This map depicts each of the Metropolitan Police divisions, arranged along borough lines, and fanning out from the centre of London. The abstract nature of each area, disassociated from a commonly understood geography, makes the map rather unique. The divisional areas are the result of intensive statistical mapping to identify problem areas and crime hotspots and allocate positions and resources to suit.

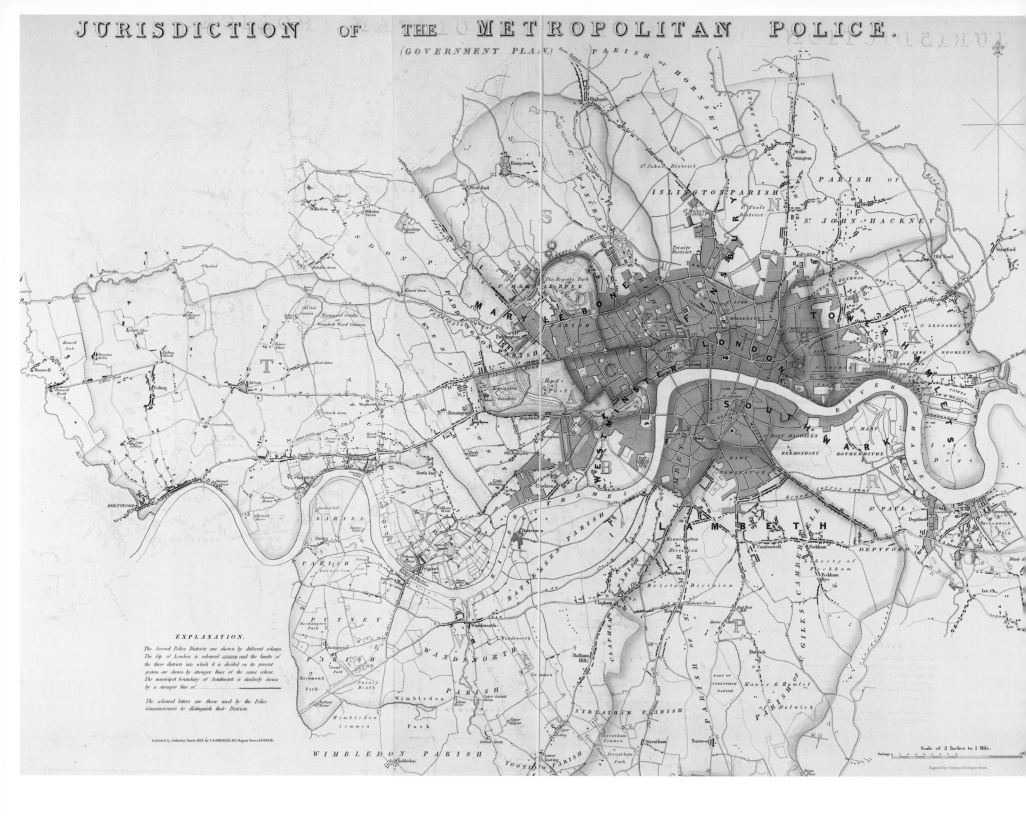

(GOVERNMENT PLAN.)

EXPLANATION.

The several Police Districts are shewn by different colours.
The City of London is coloured ——— and the limits of
the three districts into which it is divided on its present
system are shewn by stronger lines of the same colour.
The municipal boundary of Southwark is similarly shewn
by a stronger line of ———

The coloured letters are those used by the Police
Commissioners to distinguish their Districts.

Published by Authority, March 1837, by J. GARDNER, 163 Regent Street, LONDON.

Scale of 2 Inches to 1 Mile.

Jurisdiction of the Metropolitan Police, 1832–1837
Lieutenant Robert Dawson

From 1829–1839, the Metropolitan Police had authority over the area shown on this map, comprising Marylebone, Finsbury, Westminster, Tower Hamlets, Lambeth and Southwark. The City of London had its own force. The creation of a centralised force had been fiercely fought between those who believed that it should be a local concern and those, like the Home Secretary, Robert Peel, who thought the loss of liberty was outweighed by the benefits to civil order. The fear of public disorder and the insurrections in the 1820s finally swung the argument in favour of the latter constituency.

the 1830s or 1840s, there were only ten convictions for manslaughter or murder in London, and only 130 for burglary, shopbreaking, housebreaking and robbery".[12]

As if to maintain London's reputation for prostitution and murder, and to prove the scare-mongering journalists right, one set of murders of six prostitutes in Whitechapel in 1888 did upset the statistics. These became known as the Jack the Ripper killings, which obsessed the newspapers, the public and the police for months and still continue to have a powerful grip on the public imagination well over a century later. Maps were a way of keeping such stories grounded and familiar, and allowed newspapers to illustrate their copy with material that firmly located the lurid tales of crime and murder within the city that their readers lived and worked in.

The map as a means of reportage was originally developed for war reporting (for example *The Times* coverage of the Franco-German War in 1871) but was ideal for contributing factual and authoritative content to what was otherwise speculation. As a result, the fantasy of the fog-laden, dirty and violent city, and its horrific streets, became that much more real, and no doubt helped to selling more papers and newssheets as a result.

Maps that illustrate news stories are still regular features in newspapers, with London a common subject. The ready online availability of maps has made their production more straightforward but has also diminished their individual character and denuded them of the additional layers of meaning that a good map is able to convey.

The advances in mapping crime have been, on another front, chiefly the development of Geographical Information Systems (GIS) for identifying and targeting crime hot spots and allowing rapid response, and ultimately, a redesign of problem areas. Mapmakers are innovating and exploring ways of portraying complex data in visual terms so that it can tell a multiplicity of stories about places and events in an immediately readable way. The causes and symptoms of crime and anti-social behaviour are rarely attributable to a single straightforward cause and it is essential to be able to interpret the data at a multiplicity of levels if simplistic and counterproductive responses to problems are to be avoided. The ability of graphical spatial representation to achieve this, is one of the most critical and important new uses of mapping. That GIS is allowing mapmaking to be a predictive, as well as analytic, tool makes this area one of the most likely to develop in new and fascinating ways.

tle Bridge
rchy fell
ea; London
Saxons
Moon's assassin;
s with Blake asleep
obelisk; Old Street,
moor raised its twin...

Northampton Square, bought with
Masonic gold, and Bloomsbury
St George, where Hawksmoor
raised his pagan mausoleum.

Earl's Court, where
once had his well.
our stops thus far,
om scattering of

This earthbound
constellation.

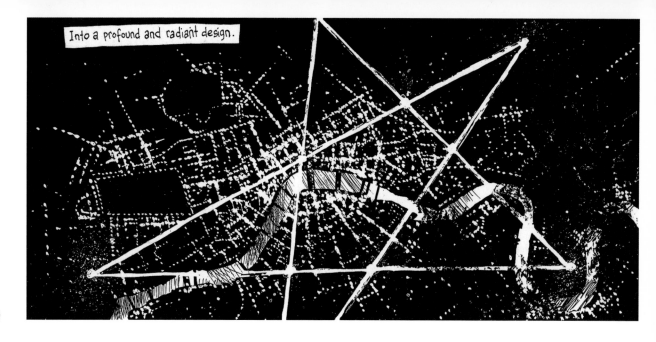

Into a profound and radiant design.

Figures 9 and 14 in *From Hell*, Alan Moore & Eddie
Campbell, 1999

The all too real mutilations and murders by Jack
the Ripper in 1888, have resulted in an immense
literature, tourist industry and general distortion of the
public image of London, and the competence of the
Metropolitan Police. *From Hell*, Moore and Campbell's
graphic novel of 1999, includes these two images
as part of its narrative—speculatively identifying
Sir William Gull, the royal surgeon, as the Ripper.
Drawn in black and white, with a grubby authenticity
appropriate to its subject, it is as much about the
contrasting wealth and lifestyles in London as it is
about the murders themselves.

© Alan Moore and Eddie Campbell.

Visiting London

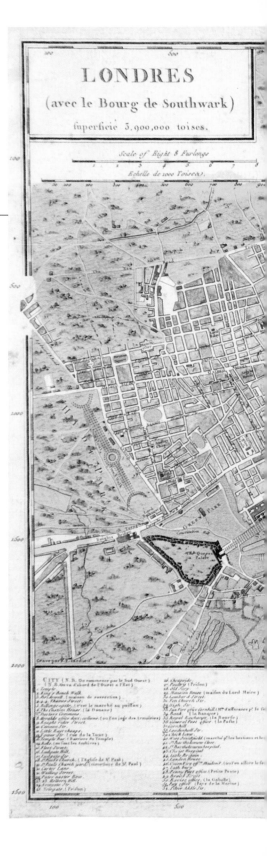

In 1841, Thomas Cook more or less invented modern tourism by arranging to take a group of 570 campaigners on a rail journey to a temperance rally in Loughborough, along a newly opened section of the Midland Counties Railway. Despite bankruptcy in 1846, his efforts brought over 165,000 people to London to visit the Great Exhibition of 1851. As a result of his activities, travel found a new purpose and a far larger clientele. Visitors to London burgeoned with the arrival of the railways in the capital, many of them with no greater intent than to explore, and to see with their own eyes, the wonders that the city possessed. Many of these visits were short and so it was important to cram as much as possible into the time available. A new type of visitor's map was necessary that could turn London into a tourist experience, in which the essential sites were linked in a non-stop itinerary. In turn, this ahistorical linkage of the key attractions invented a new narrative of London, establishing those monuments— Buckingham Palace, Piccadilly Circus, Trafalgar Square, the Houses of Parliament, St Paul's, the Tower of London, Tower Bridge and

more recently the London Eye—that have become the symbolic representations of London in the touristic imagination.

The mid-Victorian maps of London that supplied the new market of mass tourism were hardly the first aimed at travellers and visitors to London. In the sixteenth century, Francesco Velagio published an atlas of city maps, the *Raccolta di le piu*, 1595, aimed at both real and armchair travellers. More detailed maps—targeted at the limited travel industry— continued to be published at intervals over the years and also began to be included in travel books, including Joannes de Ram's map from the 1707 French guidebook *Les delices de la Grande Bretagne*. Up to 24 key sites; including Westminster Abbey and Somerset House, but also the Bethlehem Hospital and both Newgate and Bridewell prisons and Customs House, are located by a number with a key given at the bottom, a device more than familiar to us today. By 1765, another French map, *Londres avec le Bourg de Southwark,* by Pierre Francois Tardieu, had listed a total of 167 places and streets in five different areas of London as well as offering a useful translation of various words such as 'alms' (house), 'square' and 'row'.

Londres avec le Bourg de Southwark, 1765
Pierre Francois Tardieu

From Francesco Velagio's 1595 map of London onward, the visitor's view of London has been a consistent aspect of map production. This French map of 1765 is clearly designed to assist visitors with its table of principal streets and buildings, as well as its *explication de quelque mots Anglois*. This version of the map has been colour-tinted to help identify the shape and character of the city. However, it also presents a rather idealised view of a city, which is situated in forest wilderness instead of the heavily cultivated reality surrounding the urban centre.

Image courtesy of the Guildhall Library, City of London.

MOGG'S STRANGERS GUIDE TO LONDON.

Exhibiting all the various Alterations & Improvements complete to the Present Time.

Published by Edward Mogg, N°14. Great Russell Street Covent Garden.

A LIST OF 500 OF THE PRINCIPAL PLACES WITH REFERENCES TO THEIR SITUATION ON THE ABOVE PLAN.

Mogg's Strangers Guide to London, 1807–1809
Edward Mogg

Commercial mapmakers very rapidly saw the economic potential in selling maps to visitors, or 'strangers' in their terminology. But, as this early nineteenth century map shows, they did little to alter their standard product to suit such strangers. The map includes an index of streets, but this was standard practice on most maps of the period.

For the next edition, issued in 1910, the publisher changed the title to Mogg's Street Directory, though it reverted to Mogg's Stranger's Guide through London in an undated edition between 1811 and 1831. In 1838 it became Mogg's New Picture of London, a designation it kept until its final published edition in 1848.

These were essentially ordinary maps with a key added that might be useful to visitors, in much the way that Newcourt added a schedule of churches to his map of 1658, or Morgan a table of key buildings to his in 1681. The maps published in the early nineteenth century with titles such as Strangers Guide through the streets of London and Westminster, 1807, or Wallis' Guide for Strangers through London, 1841, were still straightforward street maps, (and often updated versions of previous incarnations) reprinted with additional annotation and an essential sounding title designed to grab the *ingénue* visitor.

It is only with the Great Exhibition and the arrival of mass tourism that maps showing a different version of London begin to appear. No longer are they intended to rescue the lost traveller from unfamiliar surroundings, but instead they reduce the complexity of the city to a simple set of images, and turn a visit to 'London' into a tour of the top, must-see sites along predetermined routes. It is an approach that had become familiar to the well-heeled travellers on the Grand Tour across Europe, but was a new version of 'home'. Characteristic of this new approach to mapmaking is the map produced by William Ford to coincide with the Great Exhibition. It takes the vignettes of buildings often found in the corners of maps and scatters them across a simplified London, ready for tourists to count off as they made their way to and from the show in its location of Hyde Park. It has an almost exact equivalent in today's tourist maps of the city, with their tiny drawings of the landmarks that modern visitors must fit into their busy and brief sightseeing forays.

This simplified and pictorial version of London also created another spin-off, the children's map of London with its cheerful, infantilised and cartoon view of the famous monuments and with a minimal number of streets shown. Examples include an early twentieth century nursery rhyme map with cartouches featuring rhymes relevant to various sites in London (Bartholomew, 1920) and any number of current examples (eg. the London Children's Map, 2005, designed by Guy Fox). In some cases these genres are combined for foreign school children such as the maps of Adrian McMurchie, from *Les Editions Didier*.

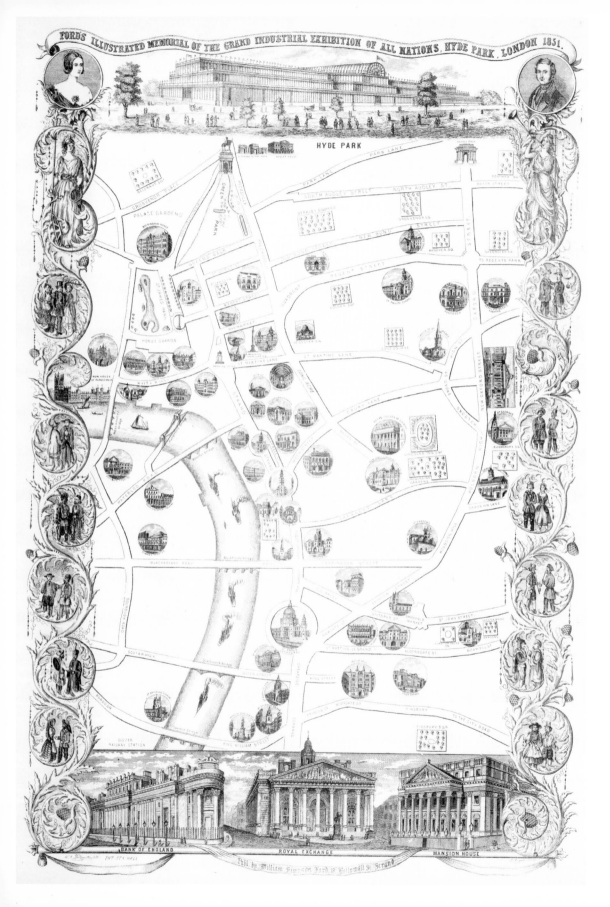

Ford's Illustrated Memorial of the Grand Industrial Exhibition of all Nations, Hyde Park, 1851
William Ford

The Great Exhibition of 1851 brought popular crowds to the capital for the first time. This commemorative map was designed to help them identify landmark places on their way through London, to and from Hyde Park. In this respect it is the first modern tourist map of the capital, eschewing a more complex understanding that captures its most essential and famous features.

Strangely, despite border illustrations of the Crystal Palace, the vignettes of Victoria and Albert and the people 'of all nations' entwined in foliage down the sides, it fails to show the location of the Exhibition itself. It equally omits other useful sites for the travelling audience, including railway stations such as Waterloo.

Right:

Wallis' Guide for Strangers through London, 1841
Edward Wallis

In Wallis's map, places of interest are heavily hatched in comparison to other buildings—a commonplace technique—but labels are hard to identify and one of the main features included, the pink-tinted boundary of the City of London, is almost of no relevance. The most appropriate aspect of the Guide is the decision to limit the area presented from west to east (Hyde Park to Whitechapel) and north to south (Regent's Park to St George's Circus). Even today, there are relatively few places that attract tourists outside this charmed area.

WALLIS'S GUIDE FOR STRANGERS THROUGH LONDON.

PUBLISHED BY E. WALLIS, 42, SKINNER STREET. AN ALPHABETICAL LIST OF 450 PRINCIPAL STREETS, WITH REFERENCES TO THEIR SITUATION. The Letters after the Names of the Streets refer to the 25 Squares on the Plan, where the streets may be found.

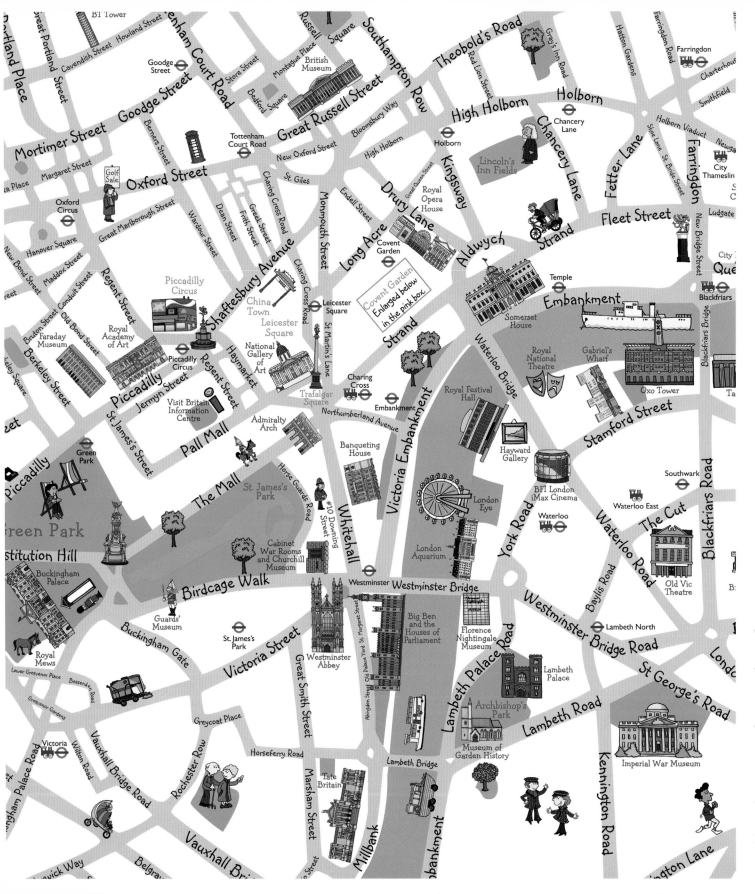

London Children's Map, 2005
Guy Fox Limited

Guy Fox's contemporary children's map of London may have its cartoon-like elements, but it is basically a straightforward map of the city that dispenses with the inessential and provides a basic map of London's streets and likely destinations. Buildings are presented as simple elevations, and a number of straightforward symbols provide a range of useful information.

Image courtesy of Guy Fox Limited.

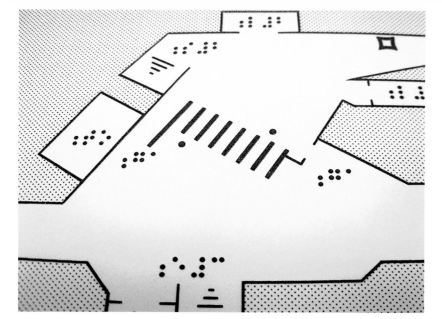

Tactile and large print maps of London
Underground Stations, 2006
Royal National Institute of Blind People (RNIB)
and London Underground

These Braille and large print Underground Station
plans were developed in a joint pilot project
between the RNIB and London Underground.
Designed in collaboration with blind and partially
sighted people, they provide pre-journey navigation
information about London Underground stations.
They are the first such maps of Tube stations ever
provided for visually impaired people.

Photo courtesy of RNIB and London Underground.

North

L2

S2

S1

EI

TO

L1

MG

MG

E1

E4

LS

5 & 6

E2

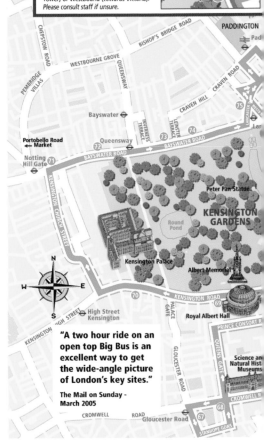

These maps make little sense of London other than to show the river and perhaps one or two parks. They certainly give little hint as to the history or character of the city or offer any opportunity for discovery. Similarly, most maps in tourist guidebooks make little effort to reveal more than landmarks or to provide street finding guides to help orientate the novice around central London (although they do give Londoners the opportunity to perceive their city through others' eyes). Looking at familiar city shapes labelled in an unfamiliar script can give the city a patina of exoticism and mystery and can sometimes reveal a different set of priorities that highlight destinations that are barely known and unexpected, even to seasoned residents.

Visitors tend to retaliate against the lack of ambition in the maps they are supplied with, by rapidly adopting the London Underground and its iconic map as their main means of navigation, with research showing that they explore London outwards from individual station nodes.[13] As a result, London is experienced and even understood by large numbers of people through the medium of one of the most geographically unreliable of mental frameworks. It is iconic maps, such as these, rather than the image in their guidebooks, that visitors have imprinted on both their souvenir t-shirts and their mind, when they return home.

"A two hour ride on an open top Big Bus is an excellent way to get the wide-angle picture of London's key sites."

The Mail on Sunday - March 2005

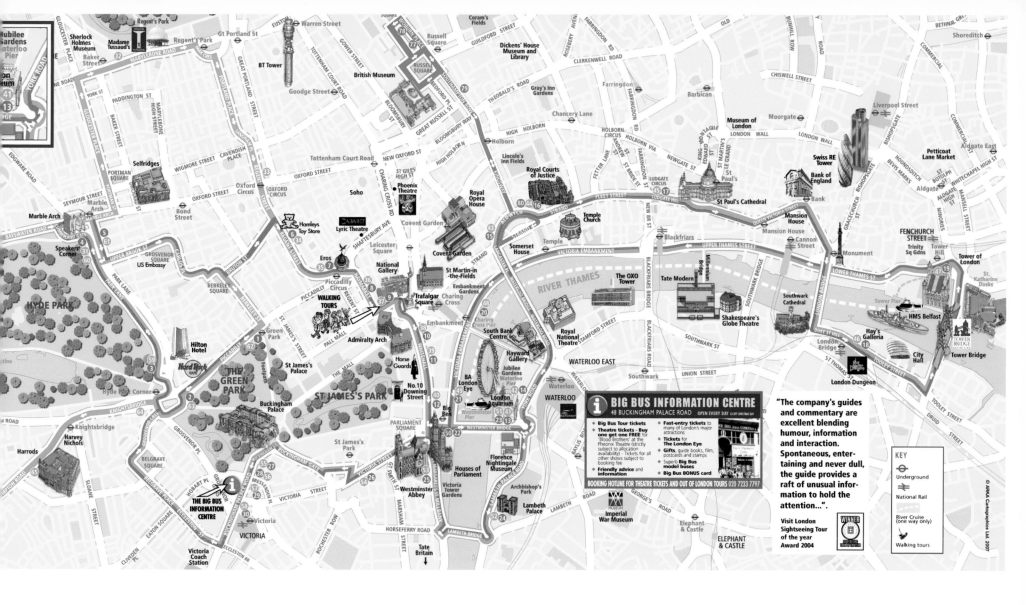

Big Bus Company Map, 2007

The demands of London's tourism industry are ever increasing. This map attempts to meet such demands and, in doing so, exceeds expectation. It shows the range of roles that London performs for tourists and a host of detail to engage and enthuse. It isn't an object of graphic beauty or elegance, but it is clear and unstinting in its efforts to put urgent information into an easily accessible format.

Image courtesy of the Big Bus Company.

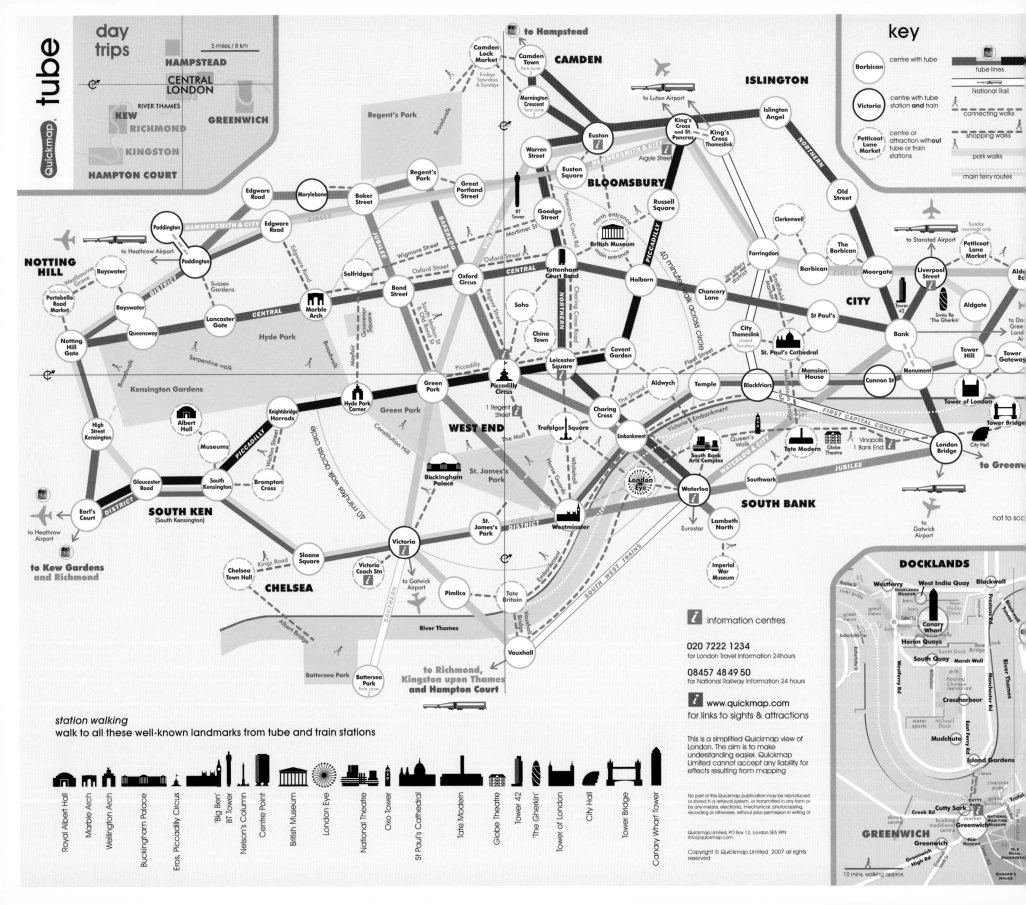

Opposite and right:

London Tube and Walk Maps, 2007

Quickmap

These maps propose using a version of the node-based London Underground system for walking, expressed as destination circles on the map with additional building silhouettes featured wherever feasible. Roads that don't assist the connections between nodes or identify attractions are commonly omitted. These maps are undoubtedly complex, both in their attempts to simplify information and to introduce many new approaches and ideas simultaneously. They could be too rich a mix to take in the popular imagination and it remains to be seen if visitors embrace these new information sources with the enthusiasm they need in order to succeed.

Image courtesy of Quickmap.

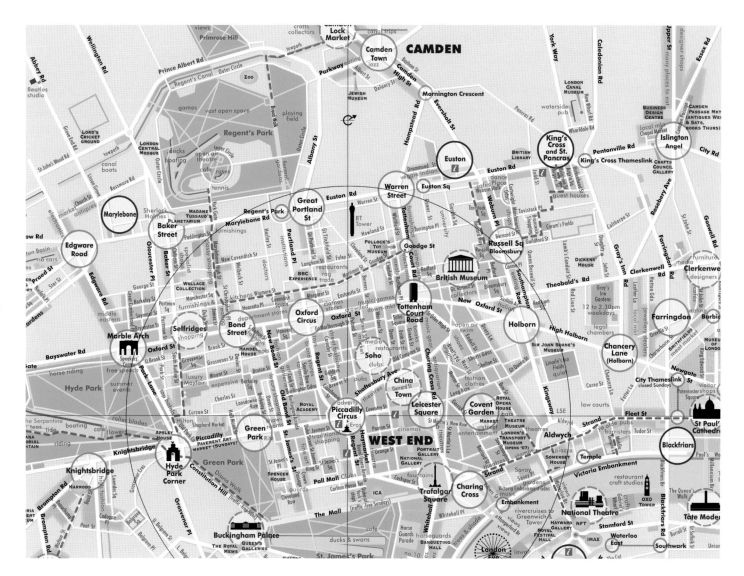

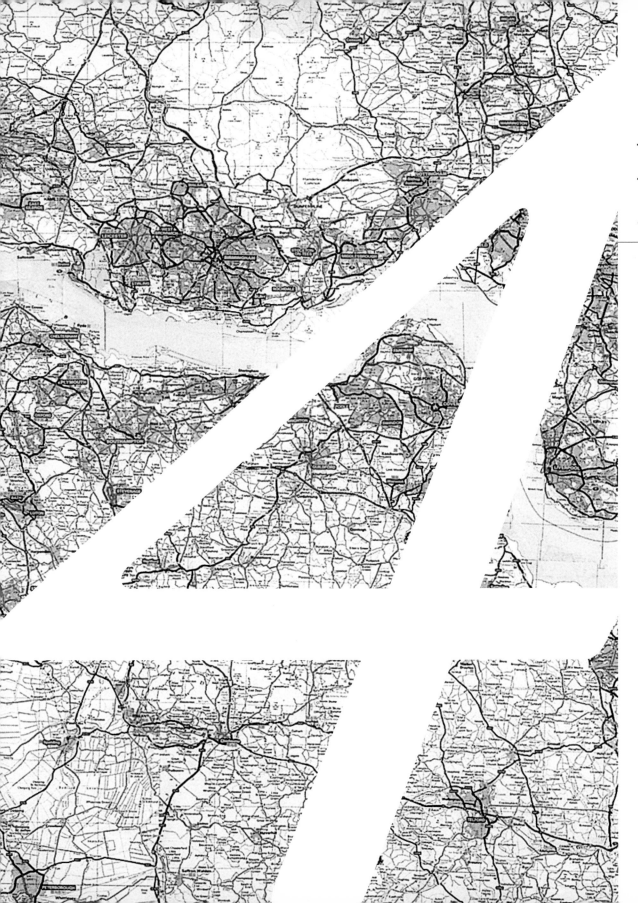

IMAGINING
LONDON

VISIONS OF LONDON

Aircastle is built on a gently sloping hill-side, and its ground-plan is practically square. It stretches from just below the top of the hill to the River Nowater, two miles away, and extends for two miles and a bit along the riverbank… The buildings are far from unimpressive, for they take the form of terraces, facing one another and running the whole length of the street. The fronts of the houses are separated by a 20 foot carriageway. Behind them is a large garden, also as long as the street itself and completely enclosed by the backs of the other streets.[1]

So starts the re-imagination of London—or perhaps it was already an established genre. Nonetheless, well before the first street maps of the city were surveyed or engraved, large-scale improvements were being envisaged to achieve Blake's much later vision of "the fields from Islington to Marylebone, to Primrose Hill and St John's Wood, were builded over with pillars of gold".[2] Mapmakers followed later, from the mid-sixteenth century on, using their particularly potent medium to rework or subvert the entire structure of the city.

Such powerful visions are more than pragmatic planning suggestions and the necessary process of improving London's street patterns and infrastructure. Visionary plans are both utopian and dystopian; with a fine and careless disregard for history, economics, individual lives and practicality. In their freedom of thought and responsibility, they are inevitably deeply tied to the moment of their conception, to the intellectual current of their time and to the impact of recent events on the nature and breadth of imagination. But, however lacking in apparent practicality such utopian schemes may appear, they must also be judged against wildly ambitious projects, such as the cutting of railway lines into and under the city, the Crystal Palace or the creation of a new city of towers on the quaysides on the West India Docks—all of which were realised—or those such as an elevated railway, intended to run the length of the Thames from London Bridge to Westminster, or the circular airport over King's Cross Station—never built, but proposed in all seriousness. London's visionaries have come in all shades of realism.

At the practical and visionary, but ignored, end of the scale is John Martin whose proposals for London's drainage problems

Left and following page:

The Strip, Exodus or the Voluntary Prisoners of Architecture from Delirious New York: A Retroactive Manifesto for Manhattan, 1975 Rem Koolhaas, Elia Zenghelis, Madelon Vriesendorp and Zoe Zenghelis

This is a dystopian vision for London—a barrier strip slicing across the city, dividing it completely into two. Such a representation could be many things: a satirical critique, a speculation resulting from Cold War paranoia or, quite possibly, a fascination with control and unrestricted luxury in the face of cruelty and poverty.

Koolhaas and his colleagues worked on this project while they were teaching at the Architectural Association (AA) in Bloomsbury, when extreme reactions were being sought by urban theoreticians to test their own assumptions as to the nature of cities and their future. Such speculations have fed directly into the later work of Koolhaas and his firm, the Office for Metropolitan Architecture (OMA), as they create ever more ambitious designs for cities across the world.

© Rem Koolhaas, Elia Zenghelis with Madelon Vriesendorp, Zoe Zenghelis.

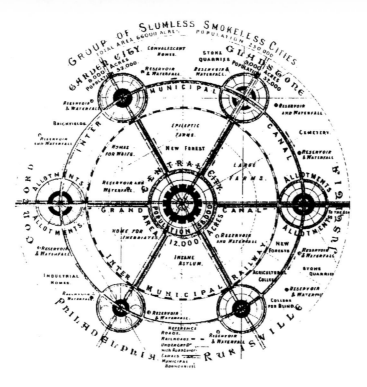

"Group of slumless, smokeless Cities"
from *Garden Cities of To-Morrow*, 1898
Ebenezer Howard

Howard's diagram of satellite garden cities around central
London is, in many ways, his version of how he would
like the map of London, and its hinterland, to look like.
His concept was extraordinarily influential on the future
of London, leading not only to the creation of his own
garden cities—Letchworth and Welwyn—but also to the
emergence of the satellite New Towns and the policy of
reducing the city's population in the 1950s and 60s.

pre-dated Bazalgette's by approximately 60 years. The Modern
Architecture Research (MARS) plan of the 1930s is possibly at
the other extreme; being derivative, impractical and highly lauded.
Emanating from the group—a collective of radical, but respected
architects and engineers, who represented the English arm of the
Congrès International d'Architecture Moderne (CIAM)—the plan
proposed completely restructuring London in accordance with the
theories of Arturo Soria Y Mata (the Spanish engineer responsible
for the concept of the linear city, who had already proposed one such
example extending from Cadiz in Spain, to St Petersburg in Russia).

The map of the ideal city has long been a fascination for
urban designers and theorists with different models, ideologies and
metaphors abounding. London, although it has been generous in
accommodating and making space for snippets of radical city ideas,
including Garden suburbs (Bedford Park, Hampstead) the Ville
Radieuse (Roehampton) and Brutalist streets in the air (for example
Robin Hood Gardens, Aylesbury Estate and Southwark) along with
a wide range of other types, has also never allowed them to spread
widely or to seriously challenge its inherent small-scale grain and

multi-centredness. London's heterogeneous mixture of styles, periods
and patterns has always maintained its dominance and no single idea
of the city has ever really been established.

Possibly for this reason, theoretical plans to transform London
are relatively rare—or perhaps the city's sheer size has allowed
experiment but discouraged attempts to establish a different orthodoxy.
In 1885, John Leighton made an intriguing stab at rationalising
London's political organisation, if not its layout on the ground,
by proposing a hexagonal structure of ward boundaries with equal
shape and area. Ebenezer Howard's diagram of a "group of slumless,
smokeless cities" of 1898 was more compelling and ultimately resulted
in not only the Green Belt around the metropolitan area but also the
New Towns, beyond the Green Belt, that inner London's population
were encouraged to leave their habitats for. The beauty of both these
proposals is their summation in simple diagrams of basic, easily
graspable principles; leaving any working out of the details to others.

Rem Koolhaas's proposals in 1992 took a linear idea even further
than the MARS Group—planting a structure east–west across the map of
London and enabling new activities as much as impeding existing ones.

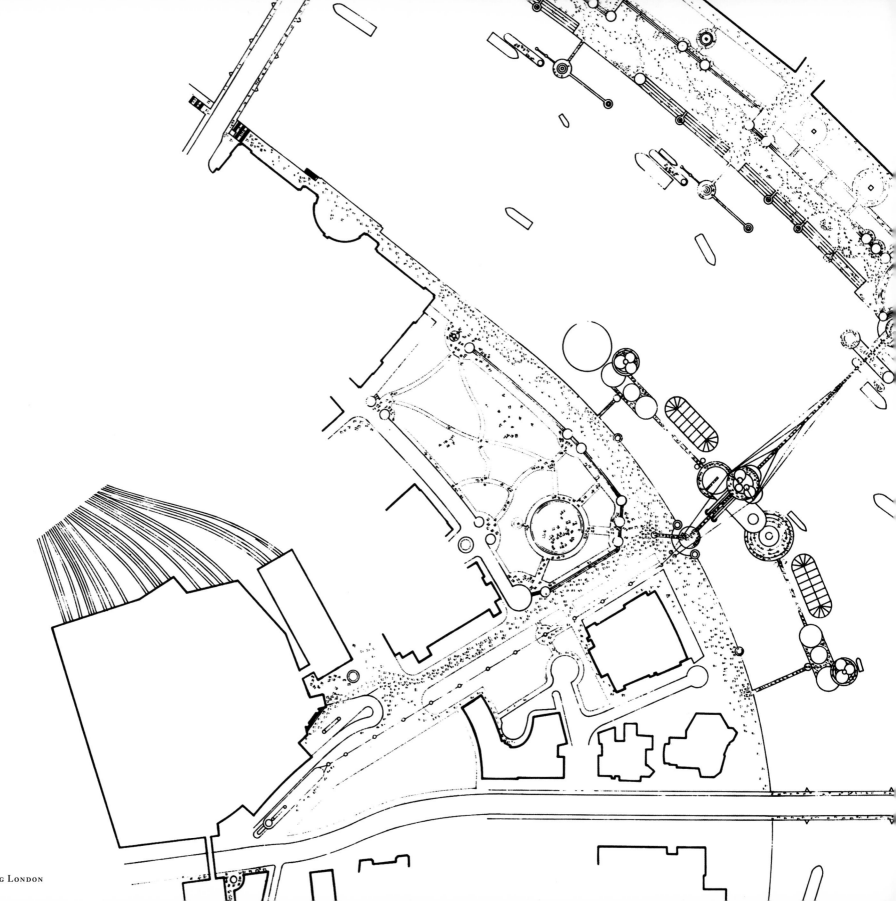

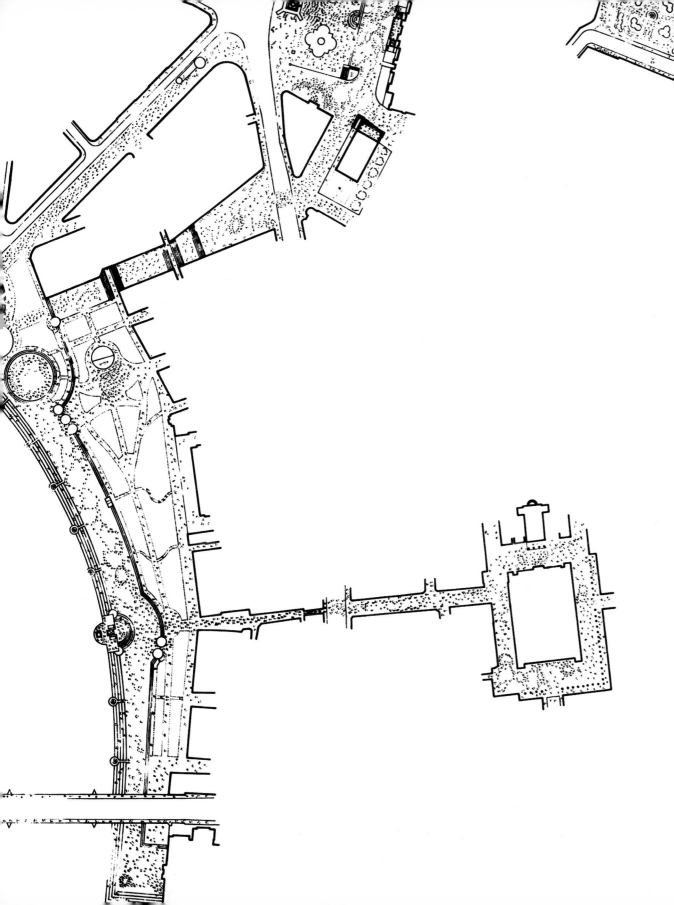

London as it could be proposal, 1986
Richard Rogers

As Chief Advisor on Architecture and Urbanism to the Mayor of London, and Chair of the Greater London Authority's Design for London Group, architect Richard Rogers has had a long engagement with the planning of London. Prepared for an exhibition at the Royal Academy (which also featured the work of Norman Foster and James Stirling), this plan proposes a new linkage of urban spaces that cross (and re-invigorate) the Thames and make pedestrian connections between Waterloo Station, through Trafalgar Square, in a northward direction. Today, many of the ideas included in this sketch have become reality.

© Richard Rogers Partnership.
Image courtesy of Rogers Stirk Harbour + Partners.

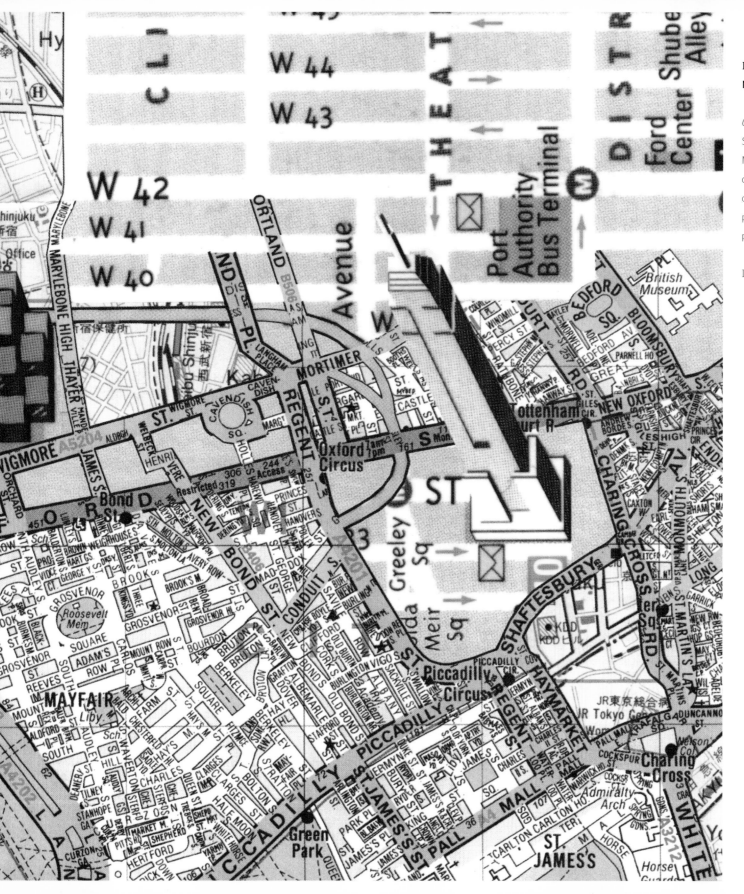

Illustration from *Guide to Ecstacity*, 2003
Nigel Coates

Guide to Ecstacity shows the global city made real.
Seven cities—Cairo, Tokyo, Rio de Janeiro, London,
New York, Mumbai and Rome—merge and interact in
delirious abandon. This map, a collage of the source
cities cited, has a close affinity to the real London—a
patchwork quilt of different approaches to urban
planning stitched together from the ragbag of history.

Image courtesy of Lawrence King.

The proposal is a provocation to those who rail against the impact of modern architecture on the city, as if to say "look what we could do if we were really trying", as well as to stimulate a re-examination of the fabric of London. The map is used to represent the status quo violated by the artist's vision. Similarly, in *Guide to Ecstacity* by Nigel Coates, 1992, the map of London is combined with those of Tokyo, Mumbai, New York, Cairo, Rome and Rio de Janeiro to create a familiar yet utterly alien Frankenstein monster of a city. It is fictional, yet recognisable; part-London, but in larger part, somewhere else altogether.

Such visions of the violently transformed city are far from the gentler proposals for re-inventing the public realm and streets of London promulgated by an older generation of architects in the early years of the twentieth century. Richard Rogers, Norman Foster, Terry Farrell and Jan Gehl have all proposed ways of rescuing London from its love affair with the car, connecting up its spaces with negotiable through-routes and new connections, especially bridges, to make this possible.

Daniel Burnham (1864–1912), the Chicago architect, famously said: "Make no little plans. They have no magic to stir men's blood and probably themselves will not be realised. Make big plans; aim high in hope and work, remembering that a noble, logical diagram once recorded will never die."[3] However, it may be the little but widespread plans, from the building out of the suburbs, to the reclaiming of streets and places, that will continue to have the biggest impact on maps of London.

Playing in the Streets

In 1935, Victor Watson and his secretary Marjorie Phillips arrived in London for the day. They had come to visit various streets and four railway stations of importance. The streets and locations they chose are a fairly random assortment of London's biggest, and most memorable, along with some relatively insignificant sites, such as Vine Street in W1, that even hardened Londoners would find hard to place. Victor Watson was the managing director of John Waddington Ltd, a firm of Leeds printers that had branched out into making cards and games, and that had just acquired the UK rights from Parker Brothers to an American game, Monopoly. In order to make the Waddington's version more popular in England, Watson decided to replace the names of Atlantic City Streets in the American original, with equivalents from London.

The version of Monopoly Watson and Phillips established that day became the standard, and probably the best-known, map of London across the world. Millions of people who have never visited London, and who might not recognise the London Underground diagram, can strongly identify with the streets of the city positioned on the board. They know that Mayfair and Park Lane are the most desirable addresses in town—positions they have maintained for centuries—and that

Whitechapel and Old Kent Road are, at a struggling best, only cheap and cheerful. It matters little that most of South London (with the exception of Old Kent Road), including any station serving the south is excluded, or that some streets are barely streets at all. The classic Monopoly game is, for very many people, their London. It has created a familiarity with the streets and with their unusual names: Piccadilly, Pall Mall, the Strand or even the Angel Islington (probably named after a cafe where Watson and Phillips stopped to eat) which, once established, is permanently engrained in memory.

Monopoly is a strange map. Many will assert that it is not a map at all but simply a diagram. However, it has a rough topological quality and is as informative about, in its way, the relationships of the streets of London as any other map. It is also a wonderful introduction to a two-dimensional representation of real places.

The use of maps in children's games didn't start with Monopoly. The Victorians published mazes based on the streets of London to entertain and instruct. There was clearly recognition in this that London's streets were a puzzle for many—not just children—but that as children apparently liked puzzles, this difficulty could at least be

Monopoly, circa 1980

The Monopoly board is one of the three most significant and enduring representations of London, all of which were created within five years of each other (the others being the London Underground diagram by Harry Beck, 1931, and the *A–Z* by Phyllis Pearsall, 1936). It still represents many strangers' mental map of London: a map that breaks many rules by its very limited depiction of the real city.

Monopoly © 2007 Hasbro. All Rights Reserved. Used with Permission.

put to some use. The challenge of navigating London streets, and later the London Underground system becomes, after this, a regular feature of board and, by the 1990s, computer games. The general consensus that the task of navigating through London is sufficiently challenging and enjoyable by itself, has allowed game inventors to freely project other narratives onto this basic premise and to explore a wide range of 'Londons' in the process. As a result, maps of London produced for both board and computer games provide a valuable lens to examine the changing popular attitudes to the city as a place and provide an insight into the concerns and prejudices of their time.

The first decade of the twentieth century produced an apparently innocent version of Ludo, called Round the Town, which involved racing across London—catching glimpses of various landmarks en-route—in order to be the first to reach the centre of town; Charing Cross. In 1940, possibly even before the Blitz had begun, children were encouraged to perceive air-raid precautions as fun, through board games such as Black-Out, 1940, where the aim was to drive various vehicles through night time streets, deliberately without benefit of landmarks. By the 1970s, these navigators transformed into taxi drivers, by way of the London Cabbie Game, the challenge of which was to make the most money possible while working the streets of London, in rivalry with one's fellow drivers. Another significant 1970s game, The London Game, was similarly themed but was otherwise very much in keeping with the spirit of Round the Town. Players use the London Underground diagram as a base, while aiming to be the first to visit six different stations, as dealt out on hidden cards. All these games have a distinct emphasis on education, but the moral improvement on offer shifts from the benign through the community-spirited to the thoroughly commercially minded—a feature also, of course, of Monopoly even though its original and ultimately ineffective intent was to illustrate and teach the evil of capitalism and the destruction it leaves in its path.

The 1980s, despite its reputation as a supremely free-enterprising decade, sees no more interest in commerce, or indeed education, and makes a determined escape into fantasy. The London games of the Thatcher years are both focused on fighting crime, with the launch of Ravensburger's still popular Scotland Yard Game, 1983, and the Sherlock Holmes Consulting Detective series set in Victorian London.

Black-Out, 1940

This game was designed and manufactured during the Second World War, quite possibly in an attempt to reiterate the need for vigilance during the Blitz. The game involves racing through the city streets at night, with delivery vehicles collecting and/or delivering a range of goods to various destinations in the West End. Cards received by the players dealt with a variety of additional wartime conditions. No landmarks are shown in this game, which both reinforces the blind journeys undertaken and suggests a degree of familiarity with the streets of inner London, which only requires an abstract map for representation.

Image courtesy of V&A Images.

WILLESDEN OLD CHURCH.

ZOOLOGICAL GARDENS.

NATIONAL GALLERY.

IMPERIAL INSTITUTE.

KENSINGTON PALACE.

TRAFALGAR SQ

HOUSES OF PARLIAMENT.

ALBERT MEMORIAL.

HARROW RD.
ELGIN AVE.
MAIDA VALE
EDGWARE ROAD
OXFORD STREET
MARYLEBONE ROAD
EUSTON
TOTTENHAM CT. RD
ST MARTINS LANE
CHARING
REGENT ST.
PICCADILLY
LADBROOKE GRO.
NOTTING HILL
HAMMERSMITH KENSINGTON KNIGHTSBRIDGE
CHISWICK
KEW BRIDGE
HAMMERSMITH BRIDGE
VICTORIA STREET
PIMLICO
CHELSEA
KINGS ROAD
YNE WALK
BATTERSEA BRIDGE
RICHMOND
FIN

GRAYS INN ROAD
OBALDS Rᴰ
CLERKENWELL Rᴰ OLD Sᵀ
HOLBORN
ST BRIDES Sᵀ
FLEET STREET
STRAND
LUDGATE HILL CHEAPSIDE ALDGATE WHITECHAPEL MILE END BOW ROAD STRATFORD
ROMAN ROAD
AW COURTS.
WATERLOO BRIDGE
WESTMINSTER BRIDGE
AMBETH RIDGE
WATERLOO ROAD
WESTMINSTER BRIDGE ROAD
KENNINGTON Rᴰ
BRIDGE Sᵀ
BLACKFRIARS BRIDGE
BLACKFRIARS Rᴰ
NEWINGTON
NEW KENT Rᴰ
LONDON BRIDGE
DEPTFORD ROAD
NEW CROSS
LEWISH
BLACKWALL TUNNEL
BLACKW TUNNEL

BRITISH MUSEUM.

BANK OF ENGLAND.

PEOPLE'S PALACE.

BLACKWALL TUNNEL.

ST. PAUL'S CATHEDRAL.

THE TOWER.

GREENWICH HOSPITAL.

An Edwardian board game in which up to eight players race from points, on the outskirts of London, into the city centre. The winner is the first to reach Trafalgar Square (here still located in Charing Cross). The playing pieces feature various modes of transport including driving (cars), cycling and walking. Numerous buildings are shown, including town halls and local landmarks such as the Archway Bridge, 1897, in Highgate and the Crystal Palace in Sydenham. The contrast between the eight radial and three orbital routes have an intriguing relationship to the ring roads later proposed by Abercrombie and others, but their use in the game is uncertain.

Image courtesy of V&A Images.

That these are also both games of strategy and skill, with role playing and problem solving to the fore, links them to the computer-based games that have followed. The maps that accompany them are heavily detailed and concerned with real places as well as the challenges of getting from one location to another, but they also lack any interpretative dimension. London is taken on its own terms and its reputation for cops and robbers appears to be assumed without further elaboration.

By the 1990s, fantasy is in full flow, and London is only available in period or supernatural garb, or both. In Tombraider III, 1998, Lara Croft visits St Paul's, a Thames-side wharf, the Natural History Museum and the abandoned Aldwych station together with various thugs and zombies. Maps are available that detail her journeys around these venues. Grand Theft Auto, a game devoted to stealing, driving and crashing cars, came to London in 1999, but with scenarios based in 1961 and 1969 and stylised maps for players to explore and exploit. The re-arrangement of London in the Grand Theft Auto map into orthogonal blocks and boroughs would have pleased many an ambitious city planner, although the equal importance given to re-spray and bomb shops and London landmarks might have given them pause for thought. A further computer game, Hellgate London, is due for current release. Set in 2038, a London in ruins is beset by the usual collection of demons, dragons and heavily armoured combatants intent on maximum despoliation. A development map shows the connections between the largely underground scenes of the game. It is intriguingly overlaid on a real map, as if geographical verisimilitude was a necessary aspect of the scenario.

In 100 years, the gamers' view of London has run from the pleasant journey through a complex transport system, through wartime resilience and crime capers to out and out apocalypse. The maps, however, tell a very different story. They have remained affable throughout, with only Grand Theft Auto showing any sign of fictional transformation. This is a scenario equally apparent from other imaginary versions of London that have associated maps; whether the 1936 Sherlock Holmes map produced to accompany the Basil Rathbone radio series in the United States, or the opening graphic to *Eastenders*, which features a digitised aerial image of London and the Thames. Fictional maps seem to find it hard to make London dirty or dangerous, unlike the many maps produced describing the all too real Whitechapel murders of Jack the Ripper, which achieve a ubiquitous grubby and unpleasant quality with no effort at all.

Scotland Yard Game, 1983
Ravensburger Spieleverlag

In Scotland Yard—a game named after the headquarters of London's Metropolitan Police—one player assumes the role of a criminal, Mr X, while the others work as a team, attempting to track him down on what is a rather complex map of London—complete with taxis, buses, waterbuses and the London Underground.

As a game that depicts London, Scotland Yard is unusual in the degree to which it encourages engagement with the city and all the complex connections between places and travel routes. Its graphics don't entirely trust players' ability to grapple with the intricacy of a true map but, nonetheless, players enthuse about its quality as an intelligent and absorbing game.

Image courtesy of Ravensburger Spieleverlag.

Grand Theft Auto—London, 1999
Rockstar North (formerly DMA Design)

The violent, destructive and criminally minded setting of the computer game, Grand Theft Auto—London, is based on a city popularised in Peter Collinson's film *The Italian Job*, 1969. The aim of the game is to climb the rung of the criminal underworld by undertaking a series of missions including bombings and assassinations. It takes place on a stylised, and not particularly accurate, map of London that is divided into segments including; Hyde Park, Camden Town, Angel, Mile End, Soho, the City, Bow, Chelsea, Westminster, Southwark, Bermondsey, Battersea, Brixton and Camberwell.

Image courtesy of Rockstar Games.

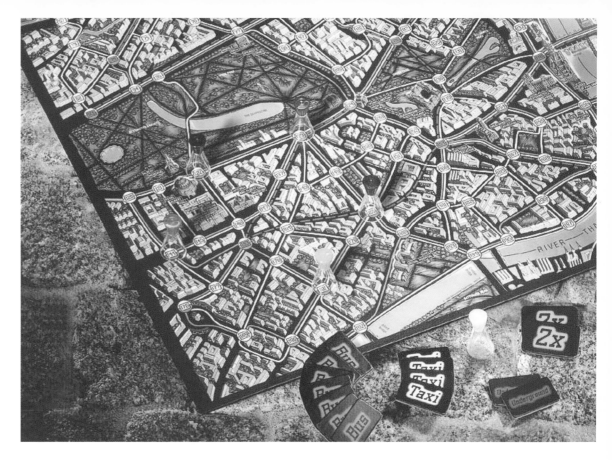

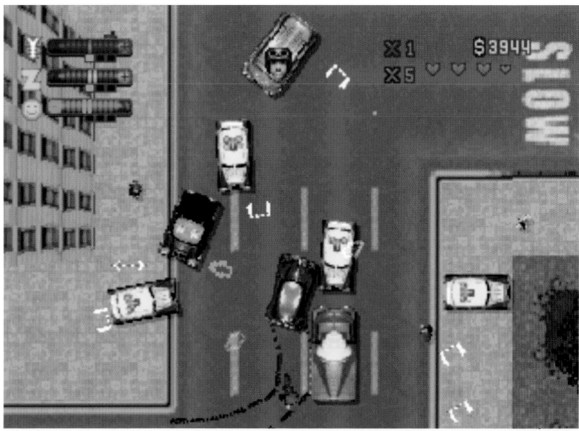

The Great Bear, 1992

Simon Patterson

Artists have long had an interest in maps but the impact of Simon Patterson's 1992 work, *The Great Bear*, on maps of London has established new territory. In his version of the London Underground diagram—printed and framed identically to the real thing—lines and stations have been renamed. *The Great Bear* has been widely published, in books, diaries and on posters, and has been extensively flattered by the imitation of others; most recently by Transport for London whose musical take on the diagram replaces lines and stations with music styles and performers/composers.

© Simon Patterson 2007
and courtesy of Haunch of Venison.

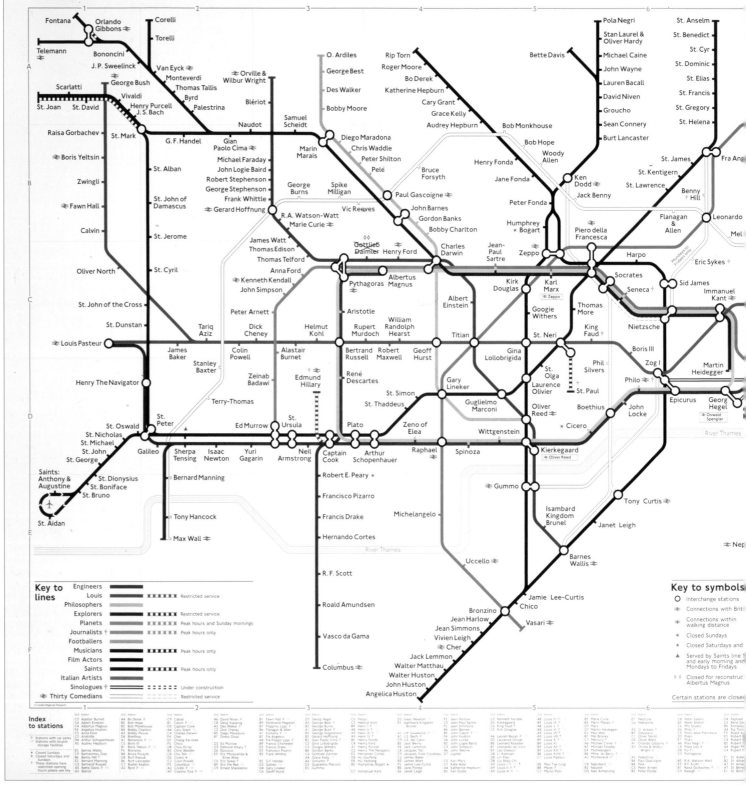

Re-interpreting the City

By the Millennium, the usual London geography, the standard cartography, had become thoroughly unreliable if not dangerously deranged. A bug had got into the system; a worm had turned. No longer were maps describing reality; reality, in the hands of Simon Patterson (*The Great Bear*, 1992) and Layla Curtis (*The Thames*, 2000), is on the blink.

The map, the serious-minded record and exploration of space and fabric, social conditions and political relationships, has been adopted by artists to investigate the emotional connections between places and people; the psychogeographical, both painful and pleasant, the dark and the light. London's psychogeographical elect have long used maps to explore and record their journeys, both real and imagined, and some, such as the walker and author Iain Sinclair, have allowed maps to define the shape of the stories they tell and exploited them so that they tell their own stories. And, as with Sinclair, or the Situationists in Paris, it is only when artists start to play with the uncomfortable relationships elucidated by the psychogeographers, that their approach begins to connect and to come alive.

Cartoonists are used to taking a sceptical view of maps and their relationship to reality. They've been prepared to twist and distort the nature of maps to represent political entities and national characteristics as well as to indulge in a light parody of pomposity and the absurd. James Gillray portrays the map of England as George III, in the guise of John Bull, shitting on the French invaders (*The French Invasion; or John Bull bombarding the bum-boats*, 1793) while Robert Dighton in *Geography Bewitched*, 1795, shows England, Ireland and Wales as Lady Hibernia submitting to John Bull. Cartoonists have kept maps in their armoury ever since; always ready to be deployed as caricatures of nationhood, personifications of evil or maidens threatened by their overpowering neighbours. Using maps as convenient shorthand, cartoonists have familiarised the concept that they are more than simply descriptive and have transformed them into a range of well-understood symbols standing in for anything from nation states to the evils of city life.

Perhaps because maps of London were constantly in flux, as they rapidly morphed and expanded, the city took time to be immortalised in caricature. It was only with the arrival of the tangle of the railways and the London Underground system that there was an image recognisable enough to be used by cartoonists.

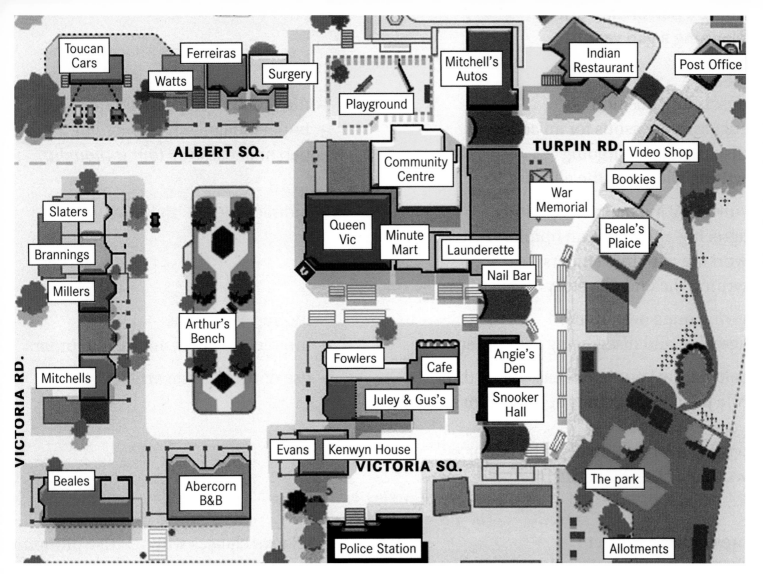

Plan of Albert Square from *Eastenders*, 1986

The imaginary realm of London takes further root with this map of Albert Square, the heart of *Eastenders*, and where all its residents live. So real is this place to many viewers that considerable efforts have been made to pinpoint it on the map of London. Possible locations include Bromley by Bow and Dalston, Hackney, as well as Stratford and Ratcliff.

Image courtesy of the BBC Picture Library.

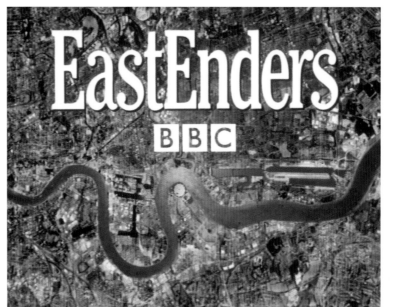

Opening graphic for *Eastenders*, 1985

The first episode of the BBC drama, *Eastenders* (set in the fictional borough of Walford, east London), was aired in 1985. Since then, each and every programme has started and finished with this familiar aerial shot over the Isle of Dogs, within the most pendulous bend of the Thames.

Image courtesy of the BBC Picture Library.

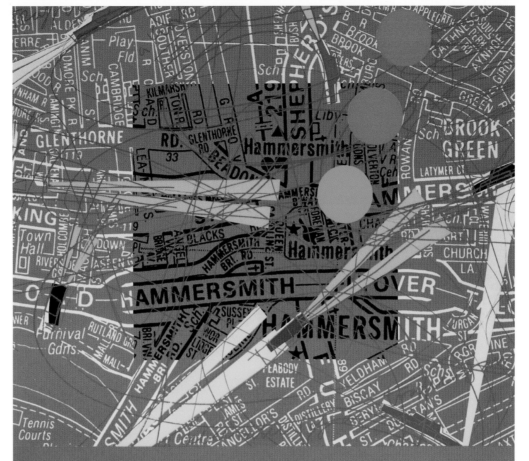

Designer Traffic Jam—Celebration of an Intersection, from *Kilometre Theatre and Seven other Seminal Architectural Projects*, 1998
Bruce McLean

Bruce McLean comments on the nature of the contemporary city, and the shared imagery of traffic flows (and chaos) evident in his own paintings and prints, in this work. The underlying map of the Hammersmith Broadway gyratory system provides a context to the work that does just what it says—celebrates the messiness of London's traffic—as well as aiming a shaft of mockery at the art world.

A designer traffic jam - a concept car park, art park pay and display, celebration of an intersection. Orchestrated bumper to bumper road rage and programmed lorry-load shedding diversion scheme on flyover, underpass and dual carriageway gridlock situation. A sleeping policeman enforcement camera, **black** spot - red routed - red light - double yellow street painting - with **white** reflector cats eyes and radar trap highway coded, conical cone line collision. A filter laned, no entry aggressive cobbled - no speed limit, no entry, congested, polluted - sin plomo'd - sans plombed, insensitive scenario for a weekend highway siren symphony, chock - a - block installation.

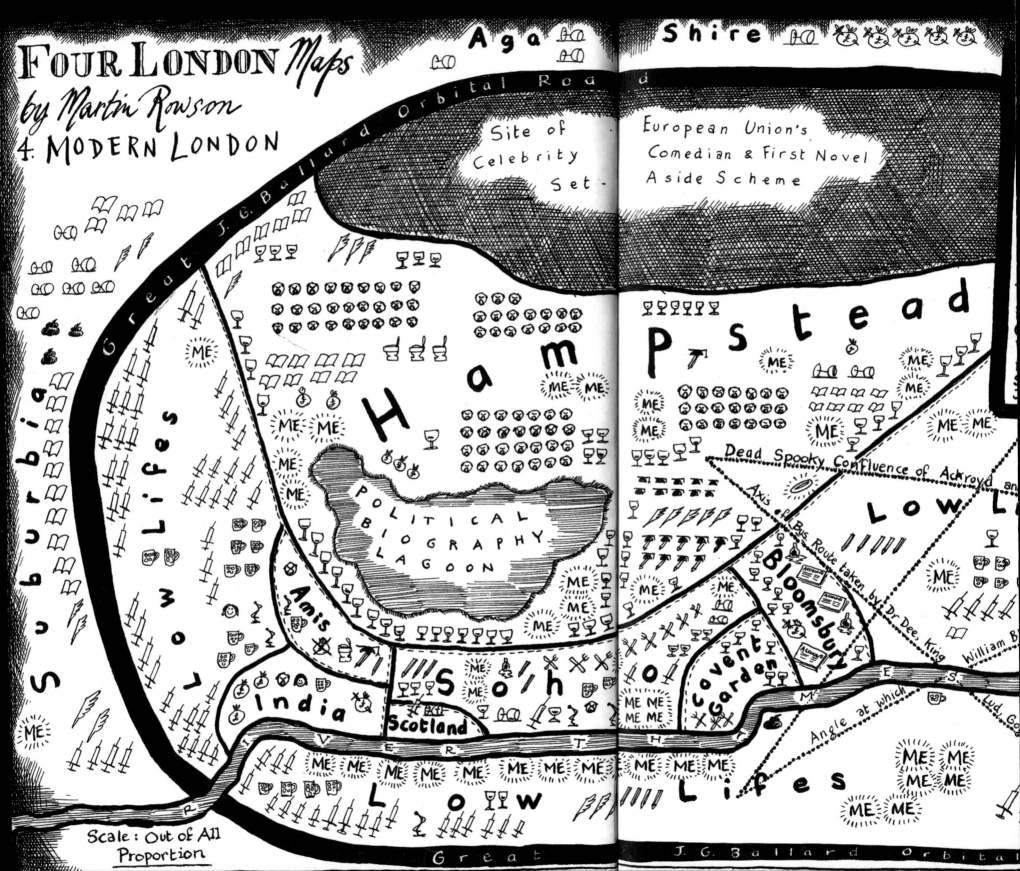

Four London Maps

by Martin Rowson

4. MODERN LONDON

Aga ... Shire

Great J.G. Ballard Orbital Road

Site of Celebrity Set-

European Union's Comedian & First Novel Aside Scheme

Suburbia

Low Lifes

ME ME ME ME ME ME ME ME ME

Hampstead

POLITICAL BIOGRAPHY LAGOON

ME ME ME

Amis

India

Scotland

Soho

Low Lifes

Dead Spooky Confluence of Ackroyd an...

Axis of Bus Route taken by Dr Dee, King

Low Li

Bloomsbury

Covent Garden

ME ME ME

William B

...E S...

Lud, Go

Angle at which

ME ME ME ME ME ME

R I V E R T H A ...

Low Lifes

ME ME ME ME

Scale: Out of All Proportion

Great J.G. Ballard Orbita...

Left, below and following pages:

Literary London: Four London Maps

from *Granta* magazine, 1999

Martin Rowson

Martin Rowson, scabrous artist and cartoonist laureate
to the Mayor of London, drew these four maps of
London in 1999 for an issue of the literary magazine,
Granta, which focused specifically on London. They
go without explanation, except for the text that
accompanies each image, and willfully distort and
adapt the map of London to Rowson's own ends.

© Martin Rowson

Image courtesy of *Granta* magazine.

FOUR LONDON MAPS
by Martin Rowson
1. CHAUCER's LONDON

As far as anyone can tell these days,
LONDON had no *literary* existence before G.CHAUCER's
"Canterbury Tales". London in those *far off days* was
very different from the bustling metropolis of
today, consisting, from what can be gleaned from
the *surviving text*, of ONE PUB and a ROAD
to CANTERBURY. There was then *no river*, nor
anything else at all, save for VAST TRACTS of
wilderness spreading *NORTH* to the Gawain
Poet's WIRRAL, *WEST* to the Malverns
and Langland's FAIR FEELD FUL OF FOLK,
and beyond there to the huge CAMELOT/
AVALON conurbation.

Quagmire of Courtly Love

SCALE: 1cm : 100 lines of Knight's Tale

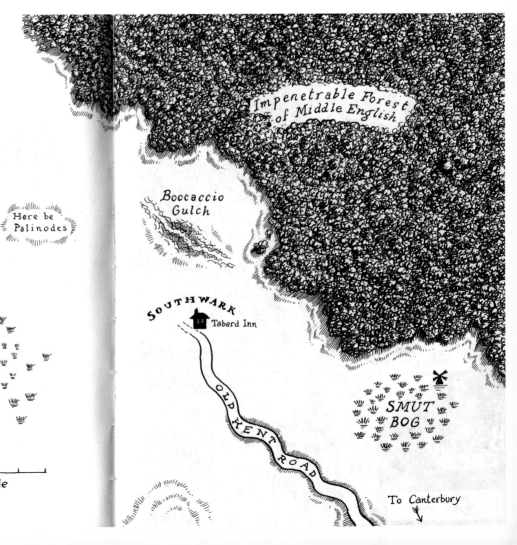

Impenetrable Forest
of Middle English

Here be
Palinodes

Boccaccio
Gulch

SOUTHWARK
Tabard Inn

OLD KENT ROAD

SMUT
BOG

To Canterbury

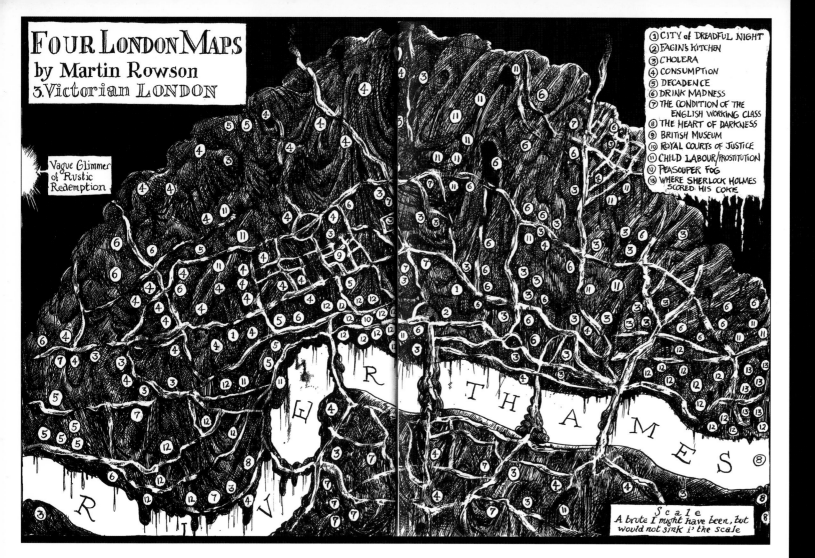

FOUR LONDON MAPS
by Martin Rowson
3. Victorian LONDON

① CITY of DREADFUL NIGHT
② FAGIN's KITCHEN
③ CHOLERA
④ CONSUMPTION
⑤ DECADENCE
⑥ DRINK MADNESS
⑦ THE CONDITION OF THE ENGLISH WORKING CLASS
⑧ THE HEART OF DARKNESS
⑨ BRITISH MUSEUM
⑩ ROYAL COURTS of JUSTICE
⑪ CHILD LABOUR/PROSTITUTION
⑫ PEASOUPER FOG
⑬ WHERE SHERLOCK HOLMES SCORED HIS COKE

Vague Glimmer of Rustic Redemption

R H THAMES R V R

Scale
A brute I might have been, but would not sink i' the scale

FOUR LO
by Martin Rov
2. 18th & Earl
COUNTRY ES

TYBURN

AIR DANCE LANE

GLO GO

RESURRECTION

FOOTPAD ALLEY

ANATOMY AVENUE

CUTTHROAT LA.
CUTPURSE ST.
DUNGHEAP LANE

GOTHICK RISE

SHEPHERDS

C

HELLFIRE C

BATH

SCALE
Ye meter generally hath 5 feet to' t.

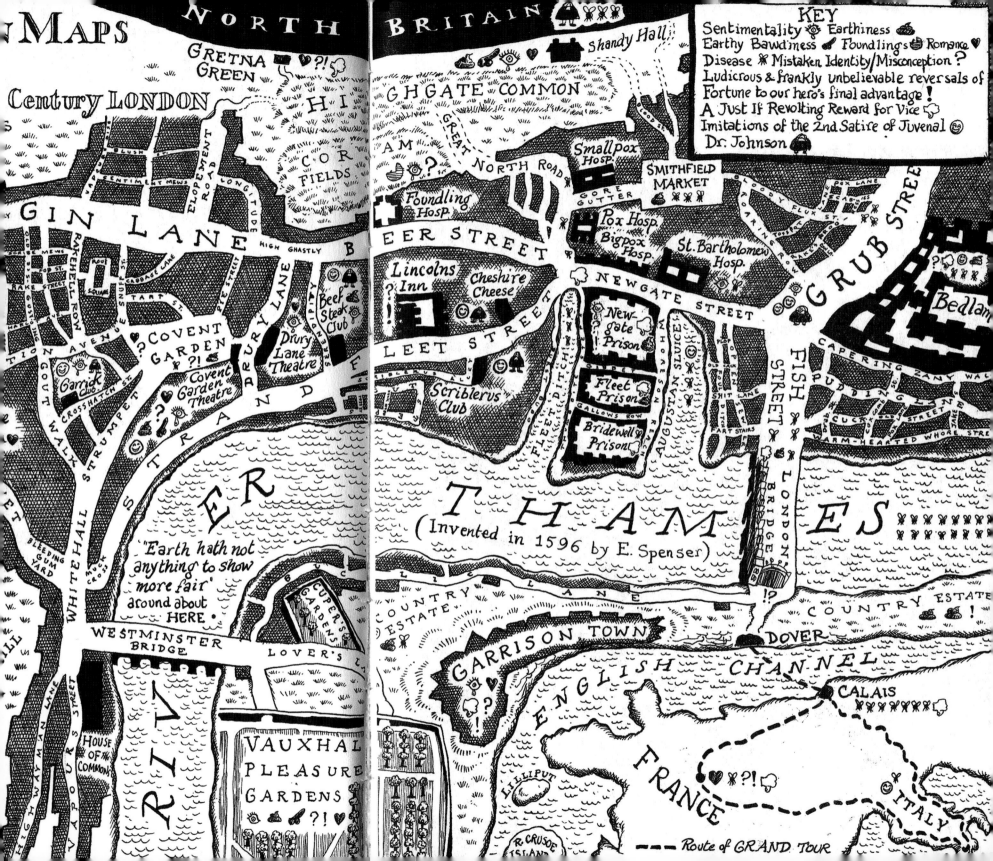

This was supplemented in the 1980s, by the nightly appearance of the opening and closing sequences of the television programme, *Eastenders*, depicting the suddenly familiar pendulous curve of the Thames around the Isle of Dogs. The Thames, as a result of *Eastenders*, became not only a literary metaphor for London and its hinterland but also a graphic one as well, and is used as such in any number of publications and other media.

Similarly, London's self-image as a patchwork of villages has proved fertile ground for cartoonists to explore maps of the city and to mock their pretensions, with the Thames commonly acting as a recognisable anchoring device. Martin Rowson's 1999 maps of literary London distort geography to reflect the caricatured mental maps of London's literati, just as Saul Steinberg's vision of New York in *View of the World from 9th Avenue*, 1976, had done before him.

By the 1990s, the map of London had become familiar enough to be available for radical re-interpretation. One such example is Simon Patterson's *The Great Bear*, a version of the London Underground diagram. Station names are replaced, line by line, by the names of engineers, philosophers, explorers, planets, journalists, footballers, musicians, film actors, saints, Italian artists, sinologues and comedians. There is no apparent logic to these substitutions but it doesn't stop the search for connections and linkages between place and personality. Such is the power of the map to convey concentrated meaning that arbitrariness cannot be allowed to stand.

In reworking the map of London, as artists including Layla Curtis and Chris Kenny have discovered, a surge of new connections and potential meanings are unleashed and allowed to expand in the viewer's mind. Imaginary London, as writers have long known and filmmakers have more recently discovered, is an exciting and extraordinary place to explore. That explorers to this shifting terrain are returning with maps, will encourage us all to pay a visit.

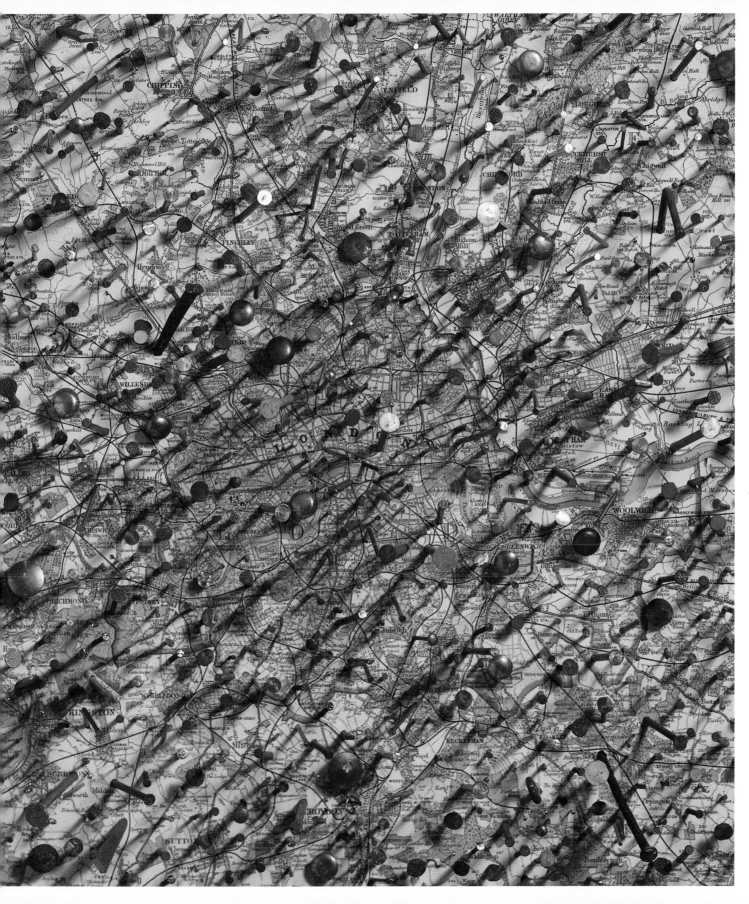

Fetish Map of London I, 2000
Chris Kenny

Kenny draws attention to the way that maps can
become fetishised objects, by creating links between
Kongo fetish figures—with their nailed in 'pledges' or
'commitments'—and the pins in a wall map. His map of
London is covered in such pins, tacks and nails to the
point of rendering it almost unintelligible.

Collection of the Museum of London.
Photograph courtesy of England & Co. Gallery, London.

Right:

An Historic Guide to Shoreditch:
How to visit the Ruins, 2001
Adam Dant

Opposite:

The Fortifications of Islington, 2006
Adam Dant

The pretensions of 'village life' in London, and the
psychological walls and boundaries erected around
its enclaves, are explored satirically by the artist
Adam Dant in these two images. In *An Historic Guide
to Shoreditch*: *How to visit the Ruins*, the Shoreditch
area (a haven of artists and creative types today)
is advanced to the year 3000, where it has been
subsumed into water and ruin.

In *The Fortifications of Islington*, the Islington area
of London can only be viewed from the outside in,
via walking along its fictional fortifications. Inside, an
almost a blank, non-descript place is depicted. All that
can be savoured otherwise are snippets of non-specific
wartime memories.

Images courtesy of Adam Dant.

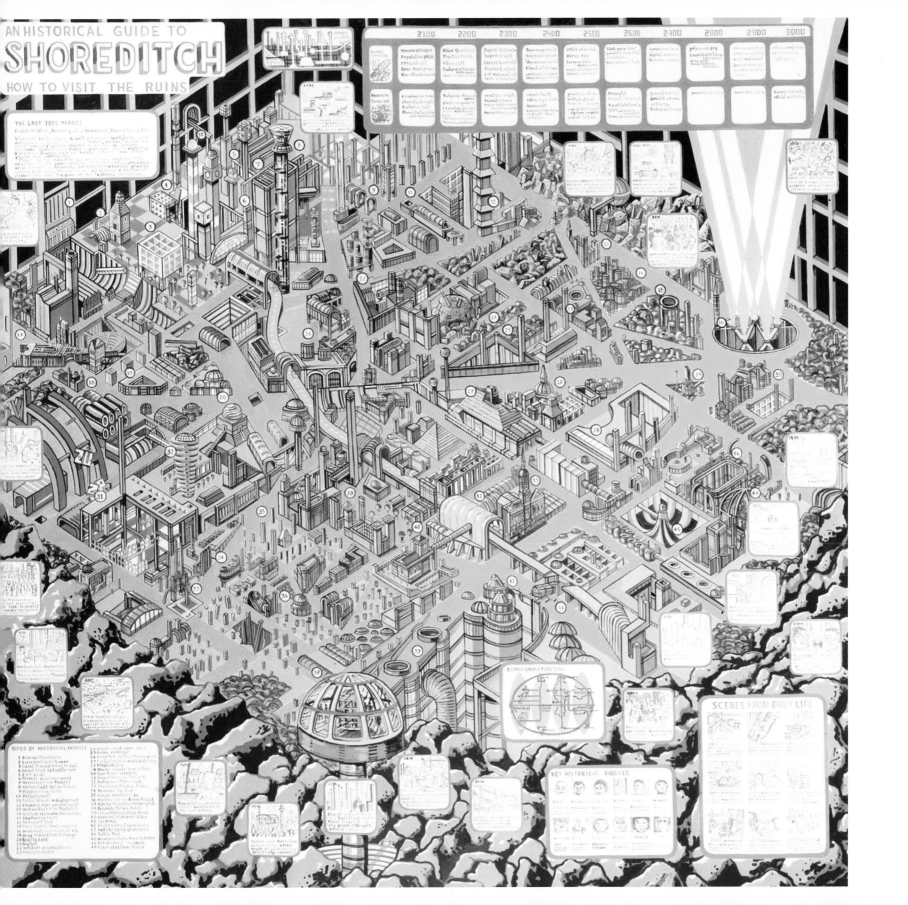

AN HISTORICAL GUIDE TO
SHOREDITCH
HOW TO VISIT THE RUINS

SCENES FROM DAILY LIFE

KEY HISTORICAL FIGURES

SITES OF HISTORICAL INTEREST

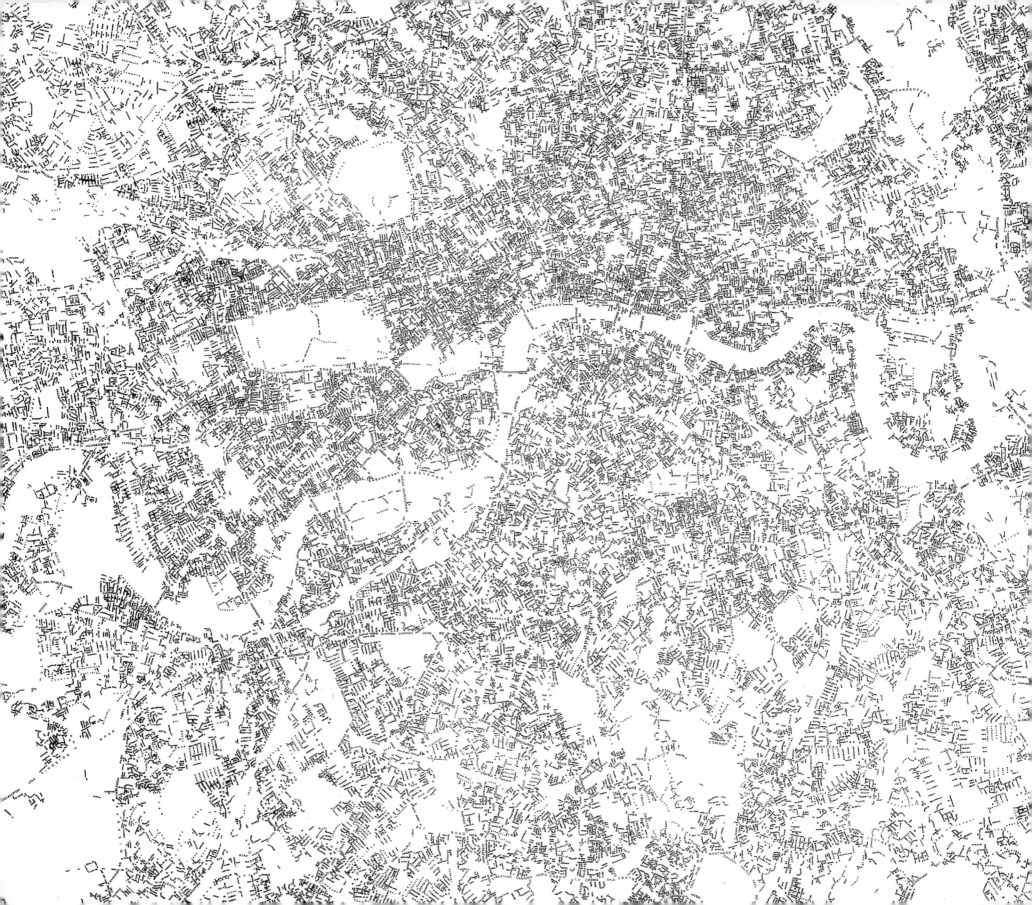

London Index Drawing, with detail, 2003
Layla Curtis

Curtis' London, for all its instant recognisability,
is a place of mysterious disappearances and
substitutions. *London Index Drawing* presents a
London that is all text and no place. Labels are all
that is left from a mass clearance of streets, the
Thames and railways. The result is a ghost town that
retains only its least tangible aspect; the arbitrary
words once associated with real places. One breath
and this gossamer of fragile words might blow away.

Images courtesy of Layla Curtis.

The Missing Voice (*Case Study B*)
from *The Walk Book*, 1999
Janet Cardiff

In this artwork by the Canadian artist, Janet Cardiff, the
audience is taken on an audio tour of London's East End,
commencing at the Whitechapel Library. The 50 minute
tour weaves through the urban landscape, providing a
fictional 'narrative' and description of the sights along the
journey. Shown here is an aerial view of the city, marking
a point on the walk when a conversation is taking place
between 'Janet' and 'George' (the artists).

© Janet Cardiff.
Image courtesy of Galerie Barbara Weiss, Berlin.

The Thames (North, South Divide), detail, 2000
Layla Curtis

In *The Thames (North, South Divide)*, a river runs
through the centre of a landscape that is clearly
London. It maintains its characteristic shape when
approaching the city centre, forcing its way through
developed areas and meandering round a heavy
drop of a peninsula. Yet this isn't London as we
know it; a quick scan of place names confirms as
such. Instead—collaged together—are the cities of
the Midlands and northern England: Manchester,
Wolverhampton, Sheffield, Liverpool and many more.
Curtis' cartography is playing with our grip on reality.

Image courtesy of Layla Curtis.

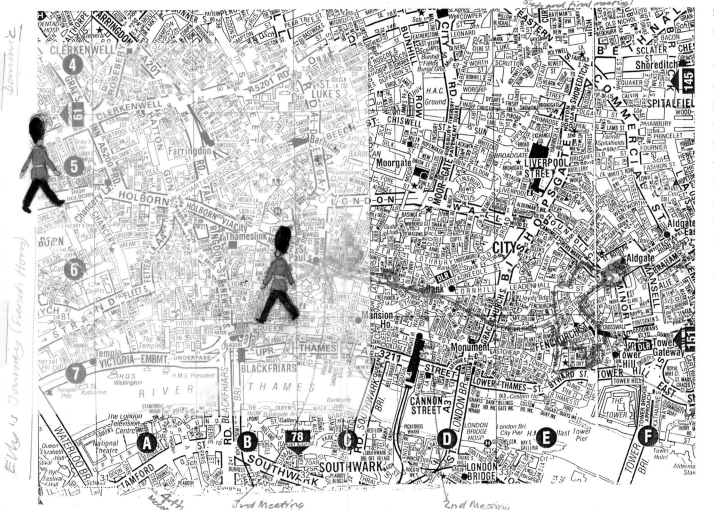

Study for *Guards*, 2003
Francis Alÿs

This recent, London-based performance follows 64 individual Coldstream Guards as they initially become lost in the city and then gradually find each other—forming back into a highly disciplined and organised troop. Alÿs, an artist with a surrealist streak who uses walking as his main art form, created a film showing the development from meander into march and so defines the shape of the streets being traversed. These studies show Alÿs' map-based work leading up to the completed film in 2005.

Image courtesy of the artist and Artangel.

VIRTUAL LONDON

He was ransacking the chaos of Stillman's movements for some glimmer of cogency…. There no longer seemed to be a question about what was happening. If he discounted the squiggles from the park, Quinn felt certain that he was looking at the letter 'E'.[4]

Every serious major company, government body and academic institute is investing in Geographic Information System (GIS) research. The big players—Google and Microsoft—seeing the strides made by the relative minnows in on-line mapping, are in costly competition to produce the ultimate two- and three-dimensional mapping applications. Such applications will allow the virtual exploration of cities, give real time advice on travel, shopping, services and entertainment and generate detailed information on the usages and needs of places both large and small.

In addition, the growing ubiquity of Global Positioning Systems (GPS) tracking is giving rise to other, alternative, maps. For example, homemade maps deriving from the invisible trails that individuals leave behind them as they move around the city or million-fold versions of Paul Auster's character, Stillman, in his short story *City of Glass*, are creating copyright free plots of the city. Open source mapping is tapping into the vast databases of information and statistics available on London and allowing individuals to upload information onto standardised geospatial frameworks and encouraging comparisons to be made between data sets. As worlds of data meet, new insights, perhaps as powerful as Dr John Snow's in Soho in 1854, for example, will emerge.

Mobile telephones, and other GPS tracking devices, provide in depth and precise information on movement and activity in cities. Face and vehicle recognition software connected to CCTV systems will supply further layers of information, as will travel cards such as London's Oyster card, and credit card purchases. All this essentially geographic data can be converted into information-rich maps that contain astonishing amounts of intelligence, useful and useless, about life lived in London. If one can interrogate the data in the right way, it may reveal almost anything about the city. These will be maps both used and abused.

Such detailed maps of real places will become ideal venues for the virtual worlds, and social and commercial networks of cyberspace,

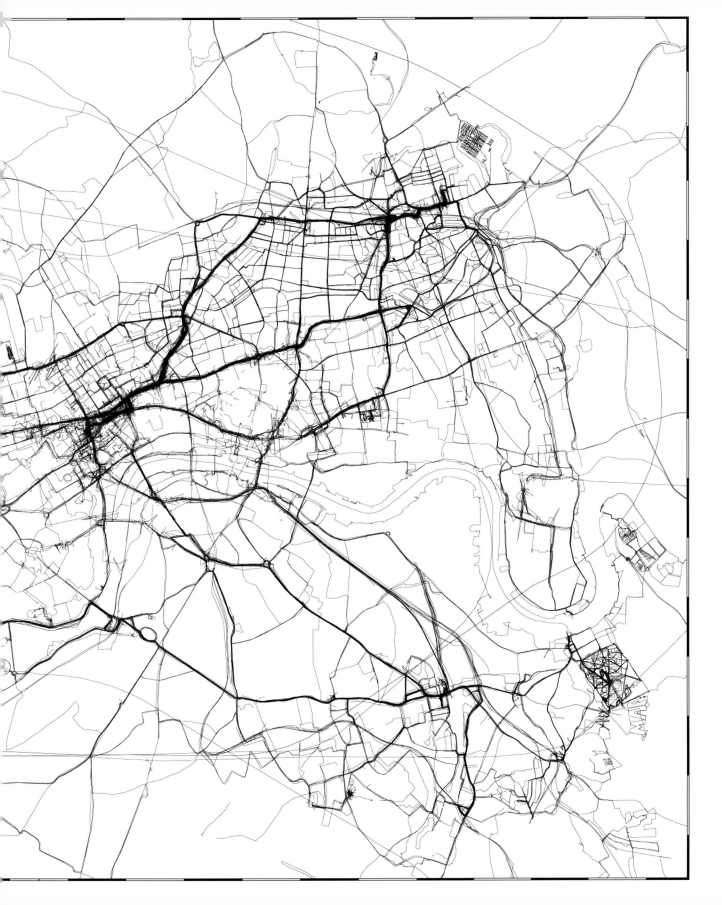

Global Positioning Systems (GPS)
Map of London, 2006
Jeremy Wood

As our lives become more closely tracked by various technologies including CCTV, number plate recognition devices and mobile phone and credit card records, we leave personal data trails behind us, crossing and crisscrossing the landscapes we inhabit. With GPS devices, these trails can be recorded and many users have taken to drawing virtual artworks and messages that only they can view. The more serious have developed into open source maps that challenge the corporate and commercial might of the mapping companies. Jeremy Wood's maps of his travels across London, over several years, are a step towards a different type of map, based on a new source of information, which also reveals the intensity of use.

Image courtesy of Jeremy Wood.

Virtual London, survey of air pollution by the Centre for Advanced Spatial Analysis, 2007

All sorts of data is collected and recorded by public agencies. This includes census information, health statistics, weather and air quality, traffic accidents, and so on. New computer programs are being developed to translate more of this information into map form, where it can be analysed far more effectively. In some cases, one set of data can be compared directly with another by linking them together on the same map. Such technology is still in its infancy but is expected to play a powerful role in future map-based research. One group working on such programs is the Centre for Advanced Spatial Analysis (CASA) at University College London, and the air pollution map shown here is one output from their experimental research. Many more will follow.

Image courtesy of CASA/University College London.

to which individuals can migrate. Their richness and historical patina will make them too tempting to ignore. Virtual avatars will soon be able to explore the streets and spaces of (almost) real places, so why not interact and experience events, both real and virtual, there as well. It is, after all, what the originals were built for. London, in the not too distant future, will be virtually rendered by three-dimensional maps, where both real and virtual lives meet, possibly in ways that are difficult to predict, and that will run the full gamut of human experience. Such renderings won't be fully accurate and may have many impossible physical attributes. They will become parallel 'Londons' capable of taking many divergent paths.

There are many virtual worlds that already have their own detailed city maps. London will soon have its own equivalents—allowing one to keep track of friends, facilitate business meetings and conferences and much more. There will be sex for sale in virtual red-light districts that could impact on their real life spatial equivalents. Complex games, far more open-ended than anything available in today's computer gaming industry, will take place on both virtual and real streets. These crossovers may change the city as we know it.

Maps are on the edge of a new phase. For the first time they are becoming inhabited. Avatars, search functions, viruses and policing systems will roam through them, both benign and malignant. It is possible that many people will spend much of their waking life navigating through virtual versions of London just as they do in the virtual worlds of Second Life or Entropia Universe today, but with the opportunity to engage in aspects of the real life of London as they do so. Maps of London may not only provide access to relevant information about places, but such places, themselves, could gradually become the focal point of the richness of activity that occurs in these maps.

Maps have always had a representational bond with the physical world and generally, so far, a one-way flow of information from reality to map. This, however, is soon to become a far more interactive relationship.

Endnotes

Introduction

[1] Amis, Martin, *London Fields*, London: Vintage, 1989, p. 367.

[2] Stephen Hall quoted in Harmon, Katherine, *You Are Here: Personal Geographies and Other Maps of the Imagination*, New York: Princeton Architectural Press, 2004. p. 17.

1 London: Growth and Change

[1] Pepys, Samuel, *Memoirs of Samuel Pepys*, 22 November 1666 entry, London: H Colburn, 1825

[2] Italian architect and surveyor Giambattista Nolli (1701–1756).

[3] James, Henry, *The Complete Notebooks of Henry James 1878–1911*, New York and Oxford: Oxford University Press, 1987.

[4] Mayhew, Henry, *The Criminal Prisons of London*, London: Griffin, Bohn, and Company, 1862.

[5] Booth, Charles, *Labour and Life of the People*, London: Williams and Norgate, 1889.

[6] Shepherd, John, Trevor Lee and John Westaway, *A Social Atlas of London*, Oxford: Clarendon Press, 1974.

[7] Howard, Luke, *The Climate of London*, London: unknown binding, 1818.

[8] Evelyn, John, *Fumifugium*, London: Gabriel Bedel and Thomas Collins, 1661.

2 Serving the City

[1] Travers, Tony, *The Politics of London*, Basingstoke: Palgrave Macmillan, 2004, p. 23.

[2] Quoted by Halliday, Stephen, in *The Great Stink of London*, Stroud: Sutton Publishing, 1999.

[3] *The Times*, 1885, quoted in Halliday, Stephen, *The Great Stink of London*, p. 58.

[4] Hyde, Ralph, quoting MBW from printed minutes, 1856.

[5] Maitland, William, *The History of London from its Foundation by the Romans to the Present Time*, London, 1739.

[6] Letter from Colonel Winterbottom, on behalf of the Director of Military Operations and Intelligence, the War Office to Director of the Ordnance Survey, 28 June 1928, in London Topographical Record, vol. 27, 1995.

[7] Letter from Colonel Winterbottom, 28 June 1928.

[8] Quoted in Halliday, Stephen, *The Great Stink of London*, p. 31.

[9] Stow, John, *A Survey of London*, London: John Windet, 1603.

[10] Inwood, Stephen, *The History of London*, London: Macmillan, 1998, p. 121.

[11] Porter, Stephen, *The Great Fire of London*, Stroud: Sutton Publishing, 1996, p. 71.

[12] Ackroyd, Peter, *London: The Biography,* London: Chatto and Windus, 2000, p. 219.

[13] Inwood, Stephen, *City of Cities*, London: Macmillan, 2005, p. 284.

[14] Gwynn, John, *London and Westminster Improved*, London, 1766, p. 7.

[15] Gwynn, John, *London and Westminster Improved*, 1776.

[16] Quoted in Barker, Felix, and Ralph Hyde, *London as it might have been*, London: John Murray, 1982, p. 59.

[17] Quoted in Porter, Stephen, *The Great Fire of London*, Stroud: Sutton Publishing, 1996, p. 102.

[18] Ackroyd, Peter. *London: The Biography*, p. 115.

[19] Abercrombie, Patrick, *Greater London Plan*, London: HMSO, 1944, p. 7.

[20] Tubbs, Ralph, *Living in Cities*, Harmondsworth: Penguin, 1942.

[21] Buchanan, Colin, *Traffic in Towns*, Harmondsworth: Penguin, 1963, pp. 11–12.

3 Living in the City

[1] Adams, George, *Geometrical and Graphical Essays,* 1791, quoted in Thomas Milne, *Land Use Map of London and Environs in 1800*, London: London Topographical Society: G B G Bull, 1975–1976.

[2] Tufte, Edward, *The Visual Display of Quantitative Information*, Cheshire: Graphics Press, 2001.

[3] Shepherd, John, Trevor Lee and John Westaway, *A Social Atlas of London*.

[4] Shakespeare, William, from the Prologue of *Henry V*: "or may we cram within this wooden O the very casques that did afright the air at Agincourt", 1600–1623.

[5] *Select Committee on Public Walks 1833*, vol. 15, issues 5–9, quoted in Inwood, Stephen, A *History of London*, p. 666,

[6] Booth, Charles, *Life and Labour of the People in London* (17 Volumes), London: Macmillan, 1902–1903.

[7] Hall, Peter, *Cities in Civilisation*, London: Weidenfeld & Nicolson, 1998, p. 657.

[8] Preamble to The Six Acts, December 1819.

[9] von Metternich, Klemens, Secret Memorandum to Czar Alexander I, 1820.

[10] Quoted by Inwood, Stephen, *A History of London*, p. 594.

[11] Quoted in Hall, Peter, *Cities in Civilisation*, p. 660.

[12] Mayhew, Henry, *London Labour and The London Poor* 1851–1852, Harmondsworth: Penguin, 1985, p. 475.

[13] Inwood, Stephen, *A History of London*, p. 604.

[14] de Graaf, K, *Legible London: A Wayfinding Study*, London: AIG, 2006.

4 Imagining London

[1] More, Thomas, *Utopia*, P Turner trans., London: Penguin, 1965, pp. 72–73.

[2] Blake, William, "Fields from Islington to Marybone", *The Complete Poems*, London: Penguin, 1977.

[3] Moore, Charles, *Daniel H Burnham: Architect, Planner of Cities*, vol. 2, New York: Houghton Mifflin, 1921.

[4] Auster, Paul, *City of Glass*, New York: Penguin, 1987, p. 69.

LIST OF ILLUSTRATIONS

INDEX

BIBLIOGRAPHY

MAPS AND MAPPING

Barber, Peter ed., *The Map Book*, London: Weidenfeld and Nicolson, 2005.

Barker, Felix, and Peter Jackson, *The History of London in Maps*, Brighton: Guild Publishing, 1990.

Black, Jeremy, *Maps and Politics*, London: Reaktion Books, 1997.

Boyle, Lucinda, and Ralph Hyde, *A Cartographic History*, London: Countrywide Editions, 2002.

Darlington, Ida, and James Howgego, *Printed Maps of London circa 1553–1850*, London: George Philip & Son, 1964.

Davies, Andrew, *The Map of London from 1746 to the Present Day*, London: Batsford, 1987.

Delano-Smith, Catherine, and Roger Kain, *English Maps: A History*, London: The British Library, 1999.

Garland, Ken, *Mr Beck's Underground Map*, Surrey: Capital transport, 1994.

Glanville, Philippa, *London in Maps*, London: The Connoisseur, 1972.

Hyde, Ralph, *The A to Z of Restoration London*, London: London Topographical Society, 1992.

Hyde, Ralph, *The A to Z of Georgian London*, London: London Topographical Society, 1982.

Hyde, Ralph, *The A to Z of Victorian London*, London: London Topographical Society, 1987.

Hyde, Ralph, *Printed Maps of Victorian London, 1851–1900*, London: William Dawson & Sons, 1975.

Laxton, Paul, *The A to Z of Regency London*, London: London Topographical Society, 1985.

Prockter, Adrian, and Robert Taylor, *The A to Z of Elizabethan London*, London: London Topographical Society, 1979.

Saunders, Ann ed., *The A to Z of Edwardian London*, London: London Topographical Society, 2007.

Saunders, Ann ed., *London Topographical Record XXIX*, London: London Topographical Society, 2006.

Saunders, Ann ed., *London Topographical Record XXVII*, London: London Topographical Society, 1995.

van Roojin, Pepin, *The Agile Rabbit Book of Historical and Curious Maps*, London: Pepin Press, 2005.

van Swaaij, Louise, and Jean Klare, *The Atlas of Experience*, London: Bloomsbury, 2000.

MAPPING LONDON

Abercrombie, Patrick, *Greater London Plan*, London: HMSO, 1944.

Abercrombie, Patrick, and John Henry Forshaw, *Council of London Plan*, London: HMSO, 1943.

Ackroyd, Peter, *London: The Biography*, London: Chatto & Windus, 2000.

Ackroyd, Peter, *Illustrated London*, London: Chatto & Windus, 2003.

Allinson, Kenneth, *London's Contemporary Architecture*, Oxford and Boston: Architectural Press, third ed. 2003.

Barker, Felix, and Ralph Hyde, *London: As it might have been*, London: John Murray, 1982.

Mitton, Geraldine ed., *Maps of Old London*, London: Adam and Charles Black, 1908.

Braybooke, Neville, *London Green*, London: Gollancz, 1959.

Brazell, John, *London Weather*, London: HMSO, 1968.

Brimblecombe, Peter, *The Big Smoke*, London: Methuen, 1987.

Bull, Guyon Boys Garrett, *Thomas Milne's Land Use Map of London and Environs*, London: London Topographical Society, 1975–1976.

Burke, Thomas, *The Streets of London*, Batsford, 1940.

Carter, Edward, *The Future of London*, Harmondsworth: Penguin, 1962.

Chandler, Tony, *The Climate of London*, London: Hutchinson, 1965.

Clout, Hugh ed., *The Times History of London*, London: Times Books, fourth ed. 2004.

Clunn, Harold, *The Face of London*, London: Spring Books, 1950.

Crawford Snowden, W, *London 200 Years Ago*, Associated Newspapers, 1946.

Cruikshank, Dan, and Peter Wyld, *London: The Art of Georgian Building*, Architectural Press, 1975.

Day, John Robert, *The Story of London's Underground*, London: London Transport, 1963.

de Graaf, K, *Legible London: A Wayfinding Study*, London: AIG, 2006.

Dutton, Ralph, *London Homes*, London: Allan Wingate, 1952.

Dyos, Harold James, *Collins Illustrated Atlas of London (1854)*, Leicester: Leicester University Press, 1972.

Kerr, Joe, and Andrew Gibson eds., *London from Punk to Blair*, London: Reaktion Books, 2003.

Fitzgerald, Percy, *The Village London Atlas*, London: Alderman Press, 1986.

Fox, Celina, *Londoners*, London: Thames and Hudson, 1987.

Fox, Edward, *London in Peril 1665–1666*, London: Lutterworth Press, 1966.

Foxell, Michael, *London*, London: Lutterworth Press, 1965.

Galinou, Mireille, *London's Lost Map*, London: Museum of London, 1998.

Gibbs-Smith, Charles, *The Great Exhibition of 1851: A Commemorative Album*, London: HMSO, 1950.

Graham-Leigh, John, *London's Water Wars*, London: Francis Boutle, 2000.

Gwynn, John, *London and Westminster Improved*, London, 1766.

Hall, Peter, *London 2000*, London: Faber and Faber, 1963.

Halliday, Stephen, *The Great Stink of London*, Stroud: Sutton Publishing, 1996.

Hamnett, Chris, *Unequal City*, London: Routledge, 2003.

Harley, Robert, *A History of the Embankment and its Bridges*, London: Capital History, 2005.

Hayes, John, and Alice Prochaska, *London Since 1912*, London: HMSO, 1973.

Hibbert, Christopher, *London: The Biography of a City*, London: Longmans, Green & Co., 1969.

Hodson, Yolande, *Facsimile of the Ordnance Surveyor's Drawings of the London Area 1799–1808*, London: London Topographical Society, 1991.

Howard, Luke, *The Climate of London*, London: unknown binding, 1818.

Howard, Luke, *On the Modifications of Clouds*, London: Taylor, 1803.

Humphries, Steve, and John Taylor, *The Making of Modern London 1945–1985*, London: Sidgwick & Jackson, 1986.

Hunt, J, and A Scoones, *Sustainable London*, The Building Centre, 2007.

Hunt, Tristram, *Building Jerusalem*, London: Weidenfeld and Nicolson, 2004.

Inwood, Stephen, *City of Cities*, London: Macmillan, 2005.

Inwood, Stephen, *A History of London*, London: Macmillan, 1998.

Johnson, Stephen, *The Ghost Map*, London: Allen Lane, 2006.

Jones, Emrys and DJ Sinclair, *Atlas of London and the London Region*, Oxford: Pergamon Press, 1968.

Linnane, Fergus, *London: The Wicked City*, London: Robson Books, 2003.

Lobel, Mary ed., *The City of London from Prehistoric Time to c1520*, Oxford: Oxford University Press, 1989.

Lutyens, Edwin, *London Replanned*, London: Royal Academy, 1942.

Lyson, Daniel, *The Environs of London*, London: T Cadell, 1792.

Marks, Stephen Powys, *The Map of Mid-sixteenth Century London: An Investigation into the relationship between a Copper-engraved Map and its derivatives*, London: London Topographical Society, 1964.

Mayhew, Henry, *London Labour and the London Poor*, Harmondsworth: Penguin, 1985.

Mayor of London, *The Draft London Plan*, London: Greater London Authority, 2002.

Mayor of London, *The London Plan*, London: Greater London Authority, 2004.

Mayor of London, *East London Green grid, primer*, 2006.

Mitchell, Rosamond and Mary Dorothy Rose Leys, *A History of London Life*, London: Longmans, Green & Co., 1958.

Murray, Peter, *Public City: Places for People*, London: New London Architecture, 2007.

Murray, Peter, *London's Moving*, London: New London Architecture, 2007.

Nairn, Ian, *Nairn's London*, London: Penguin, 1966.

O'Connell, Sheila, *London 1753*, London: British Museum Press, 2003.

Picard, Liza, *Dr Johnson's London*, London: Weidenfeld and Nicolson, 2000.

Picard, Liza, *Elizabeth's London*, London: Weidenfeld and Nicolson, 2003.

Picard, Liza, *Restoration London*, London: Weidenfeld and Nicolson, 1997.

Porter, Stephen, *The Great Fire of London*, Gloucestershire: Sutton Publishing, 1996.

Powell, Kenneth, *New London Architecture*, London: Merrell, 2001.

Pudney, John, *London Docks*, London: Thames and Hudson, 1975.

Rasmussen, Steen Eiler, *London: The Unique City*, London: Jonathan Cape, 1937 (second ed. 1948).

Reeder, David, *Charles Booth's Descriptive Map of London Poverty*, 1889, London: London Topographical Society, 1984.

Richardson, John, *The Annals of London*, New York: Cassell & Co., 2000.

Rogers, Richard, and Mark Fisher, *A New London*, London: Penguin, 1992.

Rowlanson, Thomas, Augustus Pugin, et al, *The Microcosm of London*, London: King Penguin, 1943.

Saunders, Ann, *Regent's Park from 1086 to the Present*, Bedford College, London: 1969 (second ed. 1981).

Saunders, Ann, *The Art and Architecture of London*, London: Phaidon, 1988.

Saunders, Ann and John Schofield eds., *Tudor London: A Map and a View*, London: London Topographical Society, 2001.

Schneer, Jonathan, *The Thames: England's River*, London: Little Brown, 2005.

Searle, Ronald, and Kaye Webb, *Looking at London*, London: News Chronicle, 1953.

Shepherd, John, Trevor Lee and John Westaway, *A Social Atlas of London*, Oxford: Oxford University Press, 1974.

Sinclair, Iain, *Lights Out for the Territory*, London: Granta Books, 1997.

Smith, J, *Smith's Antiquity of London.*

Smith, Stephen, *Underground London*, London: Little Brown, 2004.

Stow, John, *A Survey of London*, London: John Windet, 1603.

Strype, John, *A Survey of the Cities of London and Westminster. by John Stow*, London: S Lyne, 1720.

Summerson, John, *Georgian London*, London: Pleiades Books, 1945.

Taylor, Eva, *An Atlas of Tudor England and Wales*, London: King Penguin, 1951.

Travers, Tony, *The Politics of London*, Basingstoke: Palgrave Macmillan, 2004.

Trease, Geoffrey, *London A Concise History*, London: Thames and Hudson, 1975.

Waller, Maureen, *1700: Scenes from London Life*, London: Hodder & Stoughton, 2000.

Webb, Aston, *London of the Future*, T Fisher Unwin, 1921.

Wedd, Kit, *Artists' London*, London: Merrell, 2001.

Weightman, Gavin, and Steve Humphries, *The Making of Modern London 1815–1914*, London: Sidgwick & Jackson, 1983.

Weinreb, Ben, and Christopher Hibbert eds., *The London Encyclopaedia*, London: Macmillan, 1983.

Wellsman, John, *London before the Fire: A Grand Panorama*, London: Sidgwick & Jackson, 1973.

Whitfield, Peter, *London: A Life in Maps*, London: The British Library, 2006.

Wilson, Anthony ed., *The Faber Book of London*, London: Faber and Faber, 1993.

Wolmar, Christian, *The Subterranean Railway*, London: Atlantic Books, 2004.

Young, Michael and Peter Willmott, *Family and Kinship in East London*, London: Routledge & Kegan Paul, 1957.

White, Charles, *The City of London: Official Guide*, Cheltenham and London: Burrow & Co., 1962.

History of London, London: John Stockdale, 1796.

London City Council, *Reconstruction in the City of London*, London: City Council, 1944.

"London, the Lives of the City", London: *Granta*, 1999.

HISTORY AND BIOGRAPHIES

Ackroyd, Peter, *Blake*, London: Sinclair-Stevenson, 1995.

Aubrey, John, *Brief Lives*, London: Penguin, 2000.

Boswell, James, *The Life of Samuel Johnson*, London and Glasgow: Blackie and Son, 1946.

Boswell, James, *The London Journal*, London: The Folio Society, 2006.

Cobbett, William, *Rural Rides*, Cobbett, 1830.

Comment, Bernard, *The Panorama*, London: Reaktion Books, 1999.

Bray, William ed., *Memoirs of John Evelyn Comprising his Diary*, Warne, 1818.

Hamblyn, Richard, *The Invention of Clouds*, New York: Farrar, Straus and Giroux, 2001.

Hartley, Sarah, *Mrs P's Journey*, London: Simon & Schuster, 2001.

Hobsbawm, Eric, *The Age of Revolution 1789–1848*, London: Weidenfeld & Nicolson, 1962.

Hobsbawm, Eric, *The Age of Capital 1848–1875*, London: Weidenfeld & Nicolson, 1975.

Hobsbawm, Eric, *The Age of Empire 1875–1914*, London: Weidenfeld & Nicolson, 1987.

Hobsbawm, Eric, *Age of Extremes: The Short Twentieth Century 1914–1991*, London: Michael Joseph, 1994.

Jardine, Lisa, *The Curious Life of Robert Hooke*, New York: Harper Collins, 2003.

Jardine, Lisa, *On a Grander Scale*, New York: Harper Collins, 2002.

Jardine, Lisa, *Ingenious Pursuits*, London: Little Brown, 1999.

Leapman, Michael, *The World for a Shilling*, London: Headline, 2001.

Pepys, Samuel, *Memoirs of Samuel Pepys*, London: H Colburn, 1825.

Robbins, Jane, *Rebel Queen*, London and New York: Simon & Schuster, 2006.

Schaffer, Frank, *The New Town Story*, London: MacGibbon & Kee, 1970.

Tindall, Gillian, *The Man who Drew London: Wenceslaus Hollar in Reality and Imagination*, London: Chatto & Windus, 2002.

Tindall, Gillian, *The House by the Thames*, London: Chatto & Windus, 2006.

Tomalin, Claire, *Samuel Pepys: The Unequalled Self*, London: Viking, 2002.

Wallinger, Mark, *State Britain*, London: Tate, 2007.

GUIDES

Banks, Francis, *The Penguin Guide to London*, London: Penguin, 1958.

Fletcher, Anne, *London Eye: The Essential Guide*, London: Harper Collins, 2000.

Hardingham, Samantha, *London: A Guide to Recent Architecture*, London: Batsfod, 2002.

Jones, Edward, and Christopher Woodward, *A Guide to the Architecture of London*, London: Weidenfeld & Nicolson, 1983.

Maxey, Dale, *Seeing London*, London: Collins, 1966.

Pick, Christopher, *Children's Guide to London*, London: Cadogan, 1985.

Saunders, Nicholas, *Alternative London*, London: Nicholas Saunders and Wildwood House, fifth ed. 1974.

South Bank Exhibition Festival of Britain Guide, London: HMSO, 1951.

On Your Bike, London Cycling Campaign, 1982.

Bike it, Redbridge Friends of the Earth Cycling Campaign, 1981.

Planning and Urbanism

Abercrombie, Peter, *Town and Country Planning*, Oxford: Oxford University Press, 1943.

Arnold, Dana, *The Metropolis and its Image*, Oxford: Blackwell, 1999.

Bell, Collin and Rose Bell, *City Fathers*, London: Barrie & Rockliff, 1969.

Buchanan, Colin, *Traffic in Towns*, Harmondsworth: Penguin, 1963.

Cullen, Gordon, *The Concise Townscape*, London: The Architectural Press, 1961.

Hall, Peter, *Cities in Civilisation*, London: Weidenfeld & Nicolson, 1998.

Hall, Peter, *Urban & Regional Planning*, Hardmondsworth: Penguin, 1975.

Harrison, Paul, *Inside the inner City*, London: Penguin, 1983.

Jacobs, Jane, *The Economy of Cities*, London: Jonathan Cape, 1970.

Jones, Emrys, *Towns & Cities*, London: Oxford University Press, 1966.

Landry, Charles, *The Creative City*, London: Earthscan, 2000.

Lynch, Kevin, *The Image of the City*, Cambridge: MIT Press, 1960.

Mitchell, William, *City of Bits*, Cambridge: MIT Press, 1995.

Reader, John, *Cities*, London: William Heinemann, 2004.

Rogers, Richard, *Towards an Urban Renaissance*, London: DETR, 1999.

Rogers, Richard, and Anne Power, *Cities for a Small Country*, London: Faber and Faber, 2000.

Rowe, Colin, and Fred Koetter, *Collage City*, Cambridge: MIT Press, 1978.

Rykwert, Joseph, *The Seduction of Place*, Oxford: Oxford University Press, 2000.

Sennett, Richard, *Flesh and Stone: The Body and the City in Western Civilisation*, London: Faber and Faber, 1994.

Sharp, Thomas, *Town Planning*, Harmondsworth: Penguin, 1940.

Tubbs, Ralph, *Living in Cities*, Harmondsworth: Penguin,1942.

Mapping and Information

Harmon, Katharine, *You are Here: Personal Geographies and other Maps of the Imagination*, Princeton Architectural Press, 2004.

Pickles, John, *A History of Spaces*, London: Routledge, 2004.

Playfair, William, *Commercial and Political Atlas*, Cambridge: Cambridge University Press, 2005.

Thrower, Norman, *Maps & Civilisation*, Chicago: The University of Chicago Press, 1972.

Tufte, Edward, *The Visual Display of Quantitative Information*, Chicago: The University of Chicago Press, 2001.

Tufte, Edward, *Visual Explanations*, Chicago: The University of Chicago Press, 1997.

Fiction

Ackroyd, Peter, *Hawksmoor*, London: Hamish Hamilton, 1985.

Amis, Martin, *London Fields*, London: Vintage, 1991.

Amis, Martin, *Money*, London: Jonathan Cape, 1984.

Bateson, Frederick Wilse ed., *William Blake: Selected Poems*, London: Heinemann, 1957.

Conrad, Joseph, *The Secret Agent*, London: J M Dent and Sons, 1907.

Dickens, Charles, *Our Mutual Friend*, London: Chapman & Hall, 1865.

Doyle, Arthur, *The Annotated Sherlock Holmes*, London: John Murray, 1968.

Duffy, Maureen, *Capital*, London: Jonathan Cape, 1975.

Eliot, George, *Daniel Deronda*, Edinburgh; London: William Blackwood and Sons, 1876.

Forster, Edward, *Howards End*, London: Edward Arnold: 1910.

Grossmith, George, *The Diary of a Nobody*, Bristol: Arrowsmith, 1892.

Hollinghurst, Alan, *The Line of Beauty*, London: Picador, 2004.

McEwan, Ian, *Saturday*, London: Jonathan Cape, 2005.

Mo, Timothy, *Sour Sweet*, London: Sphere Books, 1983.

Moorcock, Michael, *Mother London*, London: Martin Secker & Warburg, 1988.

Moore, Alan and Eddie Campbell, *From Hell*, Paddington: Eddie Campbell Comics, 1999.

More, Thomas, *Utopia*, Paul Turner trans., London: Penguin, 1965.

Murdoch, Ian, *Under the Net*, London: Chatto and Windus, 1954.

Norfolk, Lawrence, *Lempriere's Dictionary*, London: Sinclair-Stevenson, 1991.

Orwell, George, *Down and Out in Paris and London*, London: Gollancz, 1933.

Shakespeare, William, *The Works of William Shakespeare*, Shakespeare Head Press, 1934.

Sinclair, Iain, *White Chappell Scarlet Tracings*, Uppingham: Goldmark, 1987.

Sinclair, Iain, and Dave McKean, *Slow Chocolate Autopsy*, London: Phoenix House, 1997.

Smith, Zadie, *White Teeth*, London: Hamish Hamilton, 2000.

Waugh, Evelyn, *Vile Bodies*, London: Chapman & Hall, 1930.

Woolf, Virginia, *Mrs Dalloway*, London: Hogarth Press, 1925.

ACKNOWLEDGEMENTS

As a boy I occasionally accompanied my father on his research trips and lectures in connection with his two books on the history of London. I was happy to amuse myself in the Greenwich Maritime Museum or the Imperial War Museum as he worked in the libraries upstairs. But perhaps it was the elaborate charts that my father prepared for his lectures that are the ultimate inspiration for this book. Sheet by sheet, and hand drawn in marker pen, they explored the growth and development of London from Roman times up to the then present of the 1960s.

That sense of London as a living, growing place, has never left me. Nor has the sense that the city is one entity; composed of myriad elements and fragments maybe; but a single coherent body nonetheless. That sense has fuelled the desire to make London legible in all its different guises. Only maps have the complexity and the potential to show the multiple layers of meaning and interpretation to achieve this. So this, I hope, is the book I went looking for and couldn't find; the book that gathers many different versions of London in one place; to mentally explore, discover and compare.

This book is a solo endeavour but I am very grateful to all those who have offered not only encouragement but also their enthusiasm, especially those whose eyes lit up when I have told them about the project, and who felt the urgent need to expound on their favourite map, board game or theory of London. The British Library kindly arranged the exhibition, London: A Life in Maps as I was writing; facilitating easy access to a wide and fabulous range of material, and revealing the huge public enthusiasm for the subject.

Duncan McCorquodale and Blanche Craig at Black Dog Publishing have offered invaluable support and assistance, but my greatest thanks must go to my family who have had any number of interesting facts tried out on them on trips through London. They get to put up with a great deal.

Many thanks to David Hale: MAPCO: Map and Plan Collection
Online: www.archivemaps.com

Author's note:
I have followed the usual convention in describing London where 'the City' (with a capital 'C') refers to the ancient, and originally walled, 'City of London', now an administrative district. The term 'city' refers to London as a whole.

© 2007 Black Dog Publishing Limited and the author. All rights reserved.

Unit 4.4 Tea Building
56 Shoreditch High Street
London
E1 6JJ

Tel: +44 (0)20 7613 1922
Fax: +44 (0)20 7613 1944
Email: info@blackdogonline.com

www.blackdogonline.com

Editor: Blanche Craig @ bdp
Designer: Julia Trudeau @ bdp

All opinions expressed within this publication are those of the
authors and not necessarily of the publisher.

British Library Cataloguing-in-Publication Data.

A CIP record for this book is available from the British Library.

ISBN: 978 1 906155 070

Every effort has been made to trace the copyright holders, but if any have been
inadvertently overlooked the publishers will be pleased to make the necessary
arrangements at the first opportunity.

Printed in Italy.

Black Dog Publishing is an environmentally responsible company. *Mapping
London: Making Sense of the City* is printed on Chromomat, a woodfree coated
paper, chlorine and acid-free, recyclable, biodegradable, produced in factories
meeting ISO 9001 and ISO 14001 norms.

architecture art design
fashion history photography
theory and things

www.blackdogonline.com